The Ethics of the Colonization of

MARS

Principle of Continuous Improvement Volume 3

Sifwat Ali

authorHOUSE®

AuthorHouse™
1663 Liberty Drive
Bloomington, IN 47403
www.authorhouse.com
Phone: 1-800-839-8640

First published by AuthorHouse 8/17/2011

ISBN: 978-1-4634-2911-9 (sc)
ISBN: 978-1-4634-2910-2 (hc)
ISBN: 978-1-4634-2912-6 (e)

Library of Congress Control Number: 2011912731

Printed in the United States of America

Dedicated to Neil Armstrong
The first man to land on the Moon

ACKNOWLEDGMENTS

I would like to acknowledge my family; my wife Fozia, children Kashif, Sabina, Henna, and Amina; my son-in-law ZeeShaan and grandson Amin for their encouragement and support. In particular, I am grateful to Kashif for the design of the front and back covers of this book. I would also like to thank the publisher for various suggestions and critiquing the book.

CONTENTS

ABSTRACT

The "Principle of Continuous Improvement" is further examined in the context of the "Colonization of Mars". The question is simple: why should the humans colonize Mars, when their history is to attempt to systematically destroy and kill any new place colonized even on the Earth?

We have not only examined the scientific and exploratory reasons but also sought help from the scriptures to see if command(s) exist to go and colonize the cosmos in general and our solar system in particular. We have attempted to pen down chapters and verses from not only the Old and New Testaments but also from the Quran, for it being the last "revelation" in the Abrahamic tradition may be more evolved from a timeline and hence scientific perspectives.

Making Mars slowly habitable, establishing corridors of worship and burial, facing issues of keeping of the human children still human even straight walking, as evolutionary pressures increase, are examined and their solutions proposed to the best of our knowledge. A source of vast electrical energy is proposed to meet the perpetual requirements of the new Planet (see appendix III).

CHAPTER I

Establishing a Baseline

1.0 Introduction

In order to start a discussion on any new idea, or even a proposal, a baseline must be established. It is a bit like going to a psychiatrist and asking for a diagnosis on a human subject. The psychiatrist must conduct a series of tests to determine the subject's baseline condition to see how the neurons are firing. In this chapter we will attempt to baseline our knowledge of the day and see what inspires the "biological life" to explore, to plan to go, and to aspire to move to a different planet.

Let us think of a giant spaceship named "paradise". In this spaceship, there is a group of beings including man that has disobeyed the commander. The commander must evict this group from the "paradise" per the rules of the spaceship. So the commander looks for a suitable planet. Let us call this planet the "Earth". The spaceship hovers over the Earth. The commander lays down the law to this group and says: "get down". He then *foretells* them more bad news "that some of you will be the enemies of the other" (i.e. you will shed each other's blood). At the end of the verdict he gives them a little good news: "and for you, this Earth is the destination and has in it all the provisions that you will need to survive, for a time".

Well! That time is up; the time has come for the *biological life* on Earth to start a journey in the cosmos on its own strength, and the first steps have already been taken.

1.1   The Urge to Explore

The urge to explore and to multiply takes a Monarch Butterfly from Canada across the United States to Mexico, some two thousand miles. Its wings barely span a few inches and the body weighs not even a quarter of an ounce and yet it fearlessly soars across the Great Lakes and into the Great Plains facing every predator and hostile elements that are unthinkable from its point of view. The pilgrimage happens every year and the Day of Judgment arrives for a generation of the monarchs with the same frequency. There is grace and beauty in this exploration. It has in it a goal; it has in it full success; and it has new life. It has in it the beautiful death of the old and after the burial, the beginning of a new sacred mission of the next generation.

Then there is a species of fish collectively known as Salmon. They lay their eggs in freshwater streams typically at high latitudes. The eggs hatch and evolve in various forms staying from one to three years in their fresh water stream. Ah, it is estimated that only 10% of all salmon eggs survive to this stage. Then they move to an area of the water, in the direction of the ocean, which is more brackish than fresh allowing the body chemistry to change, to live in salt water. In science this process is called Osmoregulation. They then proceed to the open ocean and live there for as long as four years. They endure a dangerous predatory world and under heavy ocean pressures explore a new world. Close to the end, they mature sexually and when that happens they march to a sacred pilgrimage with the only sense of the regeneration of life and return to the fresh waters they came from. Some of these fresh water streams are as far away as a thousand miles both from the Pacific and in the Atlantic Oceans. Swimming a thousand miles against the currents under water is like flying a hundred thousand miles in the air. In moving back to the birth place, they journey upstream, continuously struggling, but never losing the urge to arrive at the spawning site, even as they sense (?) other members being devoured by a host of predators. After spawning, after completing their sacred journey, they gracefully die and the next generation takes over.

Let us now move to another part of the Planet Earth, deep down, in the Mariana Trench, roughly thirty six thousand feet deep in the Pacific Ocean; in fact so deep that if Mount Everest were to be submerged totally in it, all of roughly twenty-nine thousand feet, we will still have seven thousand feet of water left above it. At the bottom of the trench the pressure is roughly 15000 psi; and that is over one thousand times that of the normal atmospheric pressure. The "Miracle of Life" still exists without light and without the warmth of the Sun. The organism and life sustains itself from the warmth derived out of the core of the Earth. There are single-celled organisms that are thought to resemble some of the world's earliest life forms. They may be single-celled called foraminifera but there are an estimated 4,000 species "living". They inhabit a wide range of marine environments, mostly on the ocean bottom. The discovery at this depth of these foraminifera living in dirt surprised even the scientists from Japan's Agency for Marine-Earth Science and Technology (who led the exploration).

Before we baseline our own ambitions, let us look at just one historical accomplishment of man. It is now almost certain that the great pyramids at Giza Egypt were built roughly 4500 years ago. It is also now well known that the three main pyramids line up almost perfectly with the three stars in the Orion belt. Computer calculations show that almost 600,000 stone blocks were used weighing anywhere from 2 to 40 tons or more. The statistics are well known. *The interest here is that of their relationship to the heavens.* Man's ability to acquire the knowledge and then take it to a level of perfection 4500 years ago cannot be penned with justice. The art of placement of huge stones and understanding of the underlying geometry as it was practiced in ancient Egypt is truly a scientific achievement. Then to take this science wherein the structures on the earth are positioned to mirror the architecture of heaven shows an unparalleled understanding. *The length of the solar year, the mean distance to the Sun, the radius of the earth, the polar radius of the earth, even the radius of the Sun are all fairly accurately measured in these pyramids.* This is as if the engineers and the priests were trying to tell someone far beyond the solar system somewhere in the Orion Constellation hundreds of light years away of an address on a planet called Earth. There are supposedly 144,000 "casing stones". They were all polished and weighing about 15 tons each with nearly perfect right angles for all six sides. Computer calculations indicated almost 41000 casing stones were used averaging 40 tons each before the face angle was cut. These would have been a beacon *reflecting Sun's light millions of miles away.* So, if someone was looking for intelligent life using intelligent devices, will quickly learn about us. It is regrettable that the knowledge base was destroyed by the "barbarians" and it took humanity four thousand years before the Soviets (socialists and communists looking ahead of the self-proclaiming self-righteous?) sent a man in the space.

It is therefore only natural that we ask about our own baseline in the year 2010. In my mind we must set that baseline a bit higher to say the least. We would like to reach the "Heaven", (paradise?) with our own strength. We would like to reach there without being dead, without the arrival of the "Day of Judgment", and without waiting for some spiritual entity to come and tell us what to do. Mars is step one in that direction and to achieve the ultimate goal of discovering the "Heaven" must be taken. We would like to fly there regularly and learn the secrets of living forever without self-destruction and with an ever improving and an impeccable quality of life.

It is not enough to have a few satellites and a Space Station. It is not enough to have reached the moon more than forty years ago and it is certainly not advisable to sit on our laurels. Many of the fundamentals are still undecided and the pace at which we are moving is unacceptable. For example, we still measure time using planetary objects and not our own cycle of the unperturbed biological clock and so on. Much remains to be done here on the planet Earth as well. Thus as a starting point on the Earth, I am proposing a department of "Exploration" as NASA (National Aeronautics and Space Administration) must follow the Principle of Continuous Improvement[1] (defined elsewhere), and graduate to a more appropriate name and mission. In this regard appropriate legislation must be written and congress must find money even if it means eating less by the politicians; even if it means that we are forced to live on just 2000 calories a day. We must force two-thirds of our people to shed weight and obesity should be declared a crime, and it is the doctors and health officials who should lose jobs if they are unable to reverse this trend. For once, the Congress should take the leadership role as the executive branch is too busy fighting three wars and bailing out the banks. Stop all the wars, cut all the funds, and bring all the troops stationed throughout the planet; stop all foreign aid, and instead demand a contribution from these countries to return the money to be used for space exploration. Leadership must think out of the box as we are about to set sail to paradise.

The funding of the "United States Department of Exploration" USDE must be all that NASA receives and many-times more. The recent announcement by President Obama over the next six years of $6 billion increase to conduct research is a disgrace. To achieve significant funding increase, the Congress should direct the IRS to allow income tax deduction from ordinary citizens to the USDE and by law oblige the government to double the amount received from the citizens (so if I contribute five dollars to USDE, the United States treasury contributes ten dollars). The 2010 NASA budget of about 18 Billion dollars is extremely low and should be an order of magnitude higher. The cumulative effect of tax deductions, government contributions, and budget allocations should be around 180 billion dollars in the first year in today's dollar value. This is still low, but is a good start. The ultimate goal should be to have at least four percent of the GDP consistently spent on USDE. The infrastructure development must be initiated at an acceptable pace.[2] The work includes both the terrestrial and extra-terrestrial areas such as ground tracking stations and so on. It is well known that time goals

---

1   A New Calendar Based upon the Human Biological Clock, and Evolution by Sifwat Ali – 2004. Also refer to Chapter VI of the book "Islam is the future – A New Calendar for Humanity" by Sifwat Ali published by Authorhouse 2008 USA.
2   Refer to the book "President Obama and the Principle of Continuous Improvement volume II" by Sifwat Ali published by Authorhouse 2009 USA

such as proposed by President Kennedy for the lunar exploration are essential in an otherwise lethargic congressional work-schedule. This is also essential if the country has any hope of turning its new generation around to get excited by math and sciences (as it was for me - back in the sixties). I cannot emphasize enough that in the end it is a funding issue as most human endeavors are. We must be looking for excuses to fund exploration on the planet Earth and the cosmos alike. The experiments suggested in reference (1) are all on our own planet and could create tens of thousands of jobs. It is my firm belief that our way of measuring time is stinting human evolution in the biological sense. The adoption of such a system shall be crucial on Mars.

Similarly the United Nations must be enticed to establish a "Department of Exploration". All country to country wars must be stopped and money allocated to such research to bring the scientists of the developing countries at the same level as those in the US, Japan or Germany. The countries bent on conducting civil wars must be blockaded to the point that it is obvious that they are headed for self-destruction. While I am a believer in self-help, it is as though there is a conspiracy to keep the scientists in developing world under-funded far below their counterparts in the first world who themselves have little to rely on. A huge amount of money is wasted in social programs in the name of humanity worldwide to the point of being scandalous. The United Nations should earmark the same (4% of each country's GDP) for exploration designated to help the scientists of the developing countries. It should be based upon a formula that is acceptable in terms of printing currency and not on foreign aid. It is not understandable as to why capital cannot be generated when legitimate work is being done on infra-structure development and space exploration. The UN charter needs significant work, as it has become a body of autocrats and is simply thrashing. *A powerful committee on space exploration must be set up, and all the permanent members and those aspiring to be permanent such as Germany, Brazil and India should contribute a minimum of five billion dollars every year to retain their seats as permanent members. <u>This will raise forty billion dollars every year to be used only for space exploration.</u>*

1.2   The Next Destination

Reality versus Fiction is the question that comes to mind, you might say; and therefore building political will is crucial on the world scene. The exploration that continues is random to say the least and without a plan that should span two hundred years. Example: The Chinese have started their lunar program. This is insane as the man has already been to the Moon. They should join forces with the rest of the world in exploration beyond the lunar surface. It is not just the Chinese; all the countries are just as bad. Example: There are no space exploration related issues discussed in any local, national, or international elections nor are there any candidates that talk about our success or failure in the field of Exploration. There is no data presented for us "the people" to decide for ourselves as to which candidate is more in favor of setting up the infra-structure for even the next fifty years as was done in the thirties or forties in just the transport sector. *It must be shown by careful analysis that in the twenty first century, the only industry left to uplift humanity to the next plateau is planetary and space exploration and that it is the one industry that will create several hundred million jobs worldwide and other terrestrial objects.* The analysis must show that it makes economic sense to bring the people closer on the Earth and to boldly set up colonies other than on Earth. In establishing the baseline, I am reluctant to propose what is so bold, it turns into fiction. For example since Mars is very cold, we could begin by warming the planet, once we get there; and in order to do so, we may set up several nuclear powered reactors deep (two miles?) below the surface to generate electricity. We could set up a solar panel ring around the planet Mars (as I have proposed one around the equator on the surface of the Earth). There are other methods of warming the planet which are biological in nature but in the end it has to be a combination of very small yet very practical steps that will yield fast results.

However, if I could enjoin fiction and reality for a brief moment, I will invent machinery to bring to the planet Mars (or to gently have it start attracting) matter in the asteroid belt until it became as nearly massive as possible to the Earth. That might take a "few" years but it will greatly remove the burden of gravity on the human population on Mars. However, that is not possible, at least not in the next thousand years. Mars, however, is habitable as will become obvious to the reader momentarily. A bit of reality is introduced here as opposed to what is close to fiction at the moment. The distance to the Earth from the Sun in light minutes is 8.32 (a light second is the *distance* 186,322 miles – i.e. light travels 186,322 miles in one second). So the Sun-Earth distance is about 93 million miles. Similarly the average distance to Mars from the Sun is 12.7 light minutes or about 142 million miles. Thus the distance between Earth and Mars varies from 36 million miles (nearest approach) to over 250 million miles, when they are at their farthest. So if you just wanted to say hello to your mother from Mars and it was at the farthest from the Earth, it will take twenty minutes one way at the speed of light – a very long distance telephone call.

1.3   Collective Effort on a Planetary Scale

Humanity in the known history has never been able to accomplish anything, except in heroic individual and collective efforts, and we might add with significant access to collective resources. Thus the Space Shuttle project could not have been possible, if it was not for the tax-payer dollars from the United States Government. The Pyramids could not have been built without enormous resources of the Egyptian Empire. Human settlements on Mars are also a project of great historic proportions and will not be successful without a significant commitment of the resources of the people of the Planet Earth in their collective enthusiasm. It is therefore essential that a worldwide infra-structure be established to collect donations from the people who believe in the cause. By the time this book becomes popular, such a structure should be available, including royalties from the sale of this book, and the money shall then be at the disposal of USDE to spend it wisely and be audited from at least three independent accounting firms on that portion that is a donation from the people. The government allocations (for NASA) already follow strict procedures. As this book is in front of you, ask your congressman so it is possible for you to go to internet and donate five dollars or whatever you can afford. Attorneys should always be on the "USDE Exploration Board" to oversee the accounting effort of collection, distribution and compliance to the US Government (or any other government that is actively involved in the exploration projects). The board shall determine the formulae for distribution of funds to various governments worldwide. The board members especially those who are attorneys in the United States would have given their permission for their names and addresses to be published in connection with this noble effort and accessible from the website "contact" procedures, properly secured. It is obvious that I am in favor of the people the governments and the businesses giving generously for this noble cause.

You may ask why this sales pitch and what's in it for me (the author). Well if I were to say "Nothing", you the reader will simply not believe. So let me give a little explanation. Let us look to the scriptures. In the context of the Old Testament Prophets and Messengers of God, the Quran has something very profound to say: Chapter XXXVI Verses 20-21; "and there came from the uttermost part of the city a man running. He cried: O my people! Follow those who have been sent. *Follow those who ask of you no fee, and who are rightly guided*". I dare not declare myself to be a prophet or a messenger, because I am not, but I can follow in the footsteps of those prophets, in their teachings and ask nothing in return for the time and energy I have spent in writing this book. I can assure you that I am not wrongly guided either – this book is a proof of that. There are many examples from modern times, of people who do things for a noble cause and ask nothing in return.

We are essentially at the end of this chapter. *The last but not the least is to address the leadership of various religious orders* in setting up the baseline. I am requesting them as well as warning them that their help is crucial. The goal is to reach the heaven (in the sense of Cosmos) and this must mean that we shall not wait unnecessarily. Our faith shall be enhanced to include reaching the Paradise on our own strength. If you, the powerful and the thoughtful, will not help, the order that you represent and are so proud of, shall wither away following the "Principle of Continuous Improvement". None of this is easy. If it was, it would have been accomplished by now. The naysayers shall be more in the beginning, and every excuse shall be on the table, but our successes and consistency of our message will make them fall in love with our message, aspirations, goals, and dedication in achieving it. This was the way of the Prophets and all that inspired us politically. They all had noble goals, lay the foundations, and figured out a way to be successful.

It is a greater challenge for the religious community (compared to the scientists and the secularists) as the Almighty has endowed them and they possess a great motivating force; and they have always shown a way forward in the history of humanity despite differences and on occasion anxiety, even a skirmish or two. Moving forward, it is the hardest thing that the humanity is to embark upon and it has to become a "deep religious duty" to get "there" wherever there is. ***It is as if Abraham, Moses, Jesus, and Mohammad have to join forces and all the gods of yester years and their followers have agreed as well to establish a home on Mars. The entire Planet Earth must now march in unison and the only enemies are the elements of outer space.***

CHAPTER II

Ethics of Planetary Colonization

2.0   Introduction

As the humans have evolved on planet Earth, the ferocity with which they kill their own kind has continued to increase. In the beginning, the humans could only use, we are told, rudimentary stone tools to skin animals and occasionally kill each other. As we continued to acquire more intelligence, we improved our lives but at the same time invented more potent tools to the point we began using the vernacular "armament". People began to kill each other in groups and tribes began to form who prided themselves in warfare. Entire "casts" were dedicated to fighting wars. The armaments led to wars at a greater scale. There came then the point at which we invented gunpowder and eventually the atomic, hydrogen, and even cobalt bombs. Just to scare the world, we went as far as to collectively punish humans by actually using two of these atomic bombs. We are able to invent stories, fiction, to start a war, and kill and destroy the other side. A wolf, the story goes (not even hungry) saw a young lamb drinking water downstream. For sport, he wanted to kill. So went up to the lamb and said "why were you making the water dirty that

I was drinking". The lamb responded "your honor, I was drinking downstream and so not possible to dirty your water". The wolf then proceeded to kill the lamb saying, "then it must have been your mother last year". We just went to war in Iraq with a similar fictional account. There are still people among us who shamelessly defend the war against a people who never harmed us.

In Star Wars fictional story, we are told of a rebellion that is trying to destroy a "Death Star" that has destroyed an entire planet by obliterating it into pieces. While we are not worried about the Earth being blown to pieces anytime soon, *we are headed in that direction.* In fact, I am reasonably sure the Earth will ultimately recover, even if we try to obliterate it as a planet, until the Sun actually becomes a "Red Giant" billions of years down the road. It can then be assumed that life on Earth might end as we know it.

2.1   The Religious Perspective

One of the most troubling questions in my mind continues to be ethical. Do we even have the right to go beyond the Planet Earth to another terrestrial body and *colonize* it? One can simply say that our history and our genetics are wired for war and bloodshed; to declare ours what is not, and kill any life in our way. We killed an entire race (even though genetically human) when we discovered the Americas, even while claiming ourselves to be civilized, only recently; and non-humans have suffered worse. Even today, we are always looking for an excuse to go to war. World Wars I and II (fought among the civilized) jointly resulted in almost sixty million casualties. What happens if another civilization on another planet simply says "you humans do not have the passport or the visa to enter our dominion? We, of course, certainly, and vehemently argue that we have the right to defend our country if invaded by others. I am sure that similar sentiment shall exist in any intelligent being elsewhere. ***Why spread bloodshed on another part of our solar system or the galaxy?***

An argument can be made that only robots programmed for peaceful exploration be sent and we should wait until we can be genetically changed to the point of abhorring collective bloodshed. That we should wait until genuine accidents are the only source of bloodshed, and wars can be included in the commandment "thou shall not kill". My sincere recommendation will be to wait until we have acknowledged the "Principle of Continuous Improvement" (PCI henceforth). *The Principle of Continuous Improvement states that all human and non-human creations, arrangements, and inventions will either perish, rendered useless, or improve in form and function. In other words, the humans must improve their wars (or the technology of it in form and function) to the point where there is no loss of human life and all arguments can be settled by some "scientific experiment"; and the result of the experiment shall be accepted by all parties. I am not proposing this (in addition to the "moral obligations") as a condition of Martian exploration. While I am not proposing that PCI should be a pre-condition; it applies, and these experiments are themselves subjected to the Principle of Continuous Improvement.*

It will take a long time to adopt PCI possibly thousands of years. For our generation, then how do we respond to this charge of an unending streak of bloodshed since time in memorial? Should we proceed with planetary exploration? Is there anything in the "scriptures"? Are there indications in the scriptures commanding us to colonize other "earths"? As in my previous works, I openly acknowledge that I am not specifically aware of any commands in the Jewish or Christian theology that encourage (or discourage) planetary exploration. There probably are oblique references, which I am asking the scholars to point to. However, we will begin with the little I know. Let us look at Genesis (1:26) "Then God said, "let us make man in our image, in our likeness, and let them rule over the fish of the sea and the birds of the air, over the livestock, over all the *Earth*, and over all the creatures that move along the ground.[3]" I think the word "Earth" is more than just the planet Earth. Earth to me is any place that is celestial in nature. Next, the use of the word "heavens" (cosmos) is intriguing in Psalm (8: 3-6) under our feet? "When I consider your heavens, the work of your fingers, the moon and the stars, which you have set in place, what is man that you are mindful of him, the son of man that you care for him? You made him a little lower than the heavenly beings and crowned him with glory and honor. You made him ruler over the works of your hands; you put everything under his feet…" *Note that when we stand on Mars, it will be under our feet.* In the New Testament, in Romans, [1:18:20] "The wrath of God is being revealed from heaven against all the godlessness and wickedness of men who suppress the truth by their wickedness, since what may be known about God is plain to them, because God has made it plain to them. For since the creation of the world God's invisible qualities – his eternal power and divine nature – have been clearly seen, being understood from what has been made, so that men are without excuse." Clearly, as a scientist, I see that Paul is emphasizing the logic of the existence of "heavens" (cosmos) and hence the "Creator" as a given; that the truth cannot be suppressed about any and all of His creations, including man.

Also, since the Quran claims to be a continuation of God's scriptures and perhaps more scientifically evolved simply by being much more current chronologically, I have combed all of its Chapters and Verses and looked for words "Earth" (ARTH) and "Heavens" (SAMAA or the plural SAMAAWAAT) in its original Arabic. In Appendix I a list of such occurrences is provided[3] for the interested reader conducting further research. There are three types of such verses. One where the word used is "Earth", second there are verses where only "Heavens" has been used, and finally verses where both of these words have been used either as *implying unity at creation* (akin to Big Bang) or simply pushing the exploration aspect, and specific commands have been given. Please note that normal occurrences of the words Sun, the Moon, other Planets, the shooting meteorites, their structure and their composition have been looked at by the author but not in any real depth and that research has been left for others.

Some general observations are in order here before the specific verses where *exploration of the heavens and the earth are ordered.* First, the heavens and the earth were created from the same source:

---

3   I would like to thank Attorney Steve Whiting of Maine, who pointed these verses in the old and new testaments and explained the Christian point of view. Steve regularly teaches Bible and was thoughtful in his explanations. I would also like to thank Aisha Akhtar of Danbury CT in helping me with this original research and very important task of combing the Quran containing hundreds of all such verses in Arabic. See Appendix I for reference.

14

Quran (XXI-30) *"Have not those who disbelieve known that the heavens and the Earth were of one piece, then We parted them and We, made every living thing (containing molecules) of water? Will they not then believe?"* Let us look at (XI-7) "And He it is Who created *the heavens and the Earth in six periods and His Throne was upon the water - that **he might try (create) you** - which of you is best in conduct (evolution)...."* And certainly in my mind implying that evolution shall be given a chance to sustain life having created the ingredients. The form of life that adapts to the changing conditions on Earth shall prosper. Also, one group among them (the humans) may be more emanate over other life forms and so on. Further when the race of Adam was expelled from paradise (II-36) "But Satan caused them to deflect there from and expelled them from the (happy) state in which they were; and We said: fall down, one of you a foe unto the other! There shall be for you on Earth a <u>habitation and provision for a time</u>." And again in (VII-24) "He said: Go down (from hence), one of you a foe unto the other. There will be for you on Earth a <u>habitation and provision for a time</u>."

Is it time for us to think about a habitation on planets other than the Earth? Is it possible that elsewhere in the cosmos, there was a real war between the two foes ("the race of Satan" and the Race of Adam) and since the history on earth has been written by the race of Adam; Satan remains a real and spiritual foe. Is it possible that one day we shall meet the foe and another war is likely in the physical sense somewhere in the cosmos? Is it possible that we will run out of resources to sustain a quality of life we have become used to on this Earth and this war is inevitable?

Secondly, one who is perceived weak can defeat the strong. (II-51) "So they routed them by God's leave and David slew Goliath..." <u>*The Goliath(s) for us humans now are the elements: the gravity of the Martian surface, the atmosphere of the planet, lack of readily available water, and energy required getting there and in sustaining life, to make the planet Mars habitable and the list goes on. We clearly acknowledge our weaknesses and should be ready to guard against those.*</u>

We now begin examination of the word Earth, the word Heaven and the words Earth and Heavens together in the context of justifying colonization of Planets in general and Mars in particular. *We begin with the example (Chapter II Verse 30th where only the word Earth has been used); it says "Behold! Thy Lord said to the angels "I will create a **Vicegerent on Earth**". They said: "Wilt Thou place in it one who will make Mischief therein and shed blood, whilst we do celebrate Thy praises and glorify Thy holy (name)". (God) said "I know what ye know not".*

Notice that angels are a type of extremely intelligent robots and so devoid of emotion, a kind of cool being. This reminds me of (a part of) a sonnet by Shakespeare:

They that have power to hurt and will do none,
That do not do the thing they most do show,
Who, moving others, are themselves as stone,
Unmoved, cold, and to temptation slow,
They rightly do inherit heaven's graces
And husband nature's riches from expense;
They are the lords and owners of their faces,
Others but stewards of their excellence.

So when God talks of a vicegerent, the first response from the angels is cool. But God is about to make Adam (and the human race) endowed with emotions, passion, and ultimately love. It is love that can make humans excel the angels and the lack of it can also make them worse than a beast and it gives them the ability to call the shots independently when needed. Notice also, that the word "Earth" has been used once here, yet in my mind it does NOT have singular meanings of the "Planet Earth" as there is another linguistic implication in the Arabic word ARTH. It also implies Terrestrial Structure that can be divided, and ownership established; a place where there are mountains, canyons, valleys, deserts, and water and where a "Kingdom" can be established; hence the need for a Vicegerent.

Let us now take another example where the word "SAMAA" or "heavens" has been used (by now man has been "placed" on Earth).

The example concerns[4] JESUS (Chapter V verses 112 thru 114). The word "heavens" has been used once in number 112 and once in number 114. The words may be referring to the Last Supper. The verses read: *"When the disciples said: O Jesus, son of Mary! Is Thy Lord able to send down for us a Table Spread with food from heaven? He said: Observe your duty to God, if ye are true believers". (They said): We wish to eat thereof, that we may satisfy our hearts and know that thou hast spoken truth to us, and that thereof we may be witnesses. Jesus, son of Mary, said: O God, Lord of us! Send down for us a table spread with food from Heaven, for the first of us and the last of us, a solemn feast, and a sign from Thee. Give us sustenance, for Thou art the best of the Sustainers".* Two things come out of the verses quoted here. First, it is true that man has been given the genetic abilities to do mischief and other serious crimes. But that itself cannot be the reason to not "plant" him on the Earth, the most hospitable of all the locations besides the "Heaven" itself. Secondly, in His wisdom God seems to provide sustenance (from Heaven) at locations where sustenance was an issue of belief. In another verse, Moses is also provided food directly from the Heaven in the desert of the Earth. In this example as well, the location on the surface of the Earth could not support habitation; little or no water, or fertile land. A bit of a tangent here; the Christian reader may not have known of the description of this miracle of Jesus in the Quran. I would like to quote the next verse (115) as well. "God said: Lo! I send it down for you. And who so disbelieveth of you afterword, him surely will I punish with a punishment wherewith I have not punished any of (My) creatures". The implication is that if God loves you a lot and you continually disobey Him, the punishment is also very strong. It should be noted that the punishment may be purely of extreme psychological nature.

Finally let us take an example where both the words Earth and the Heavens have been used i.e. Quran (LV verses 31-33). Verse 31 says

---

4   Please note that one Arabic sentence does not always equal one English sentence. Hence there may be an occasional discrepancy in the numbering of verses in older *English translations*. It does not affect the original Arabic text. For consistency, the author has used the popular translation and numbering by Pickthall, also consistent with the "official numbering" of the Islamic Research, IFTA organization.

"we shall (momentarily) dispose of you o ye dependents (Human and Jinn)". *This means that a burden is about to be lifted by the Almighty.* Is the disposition of the burden being talked about that of the burden of the Gravitational pull? The Arabic word used to describe the two groups of creation in this verse is "Saqalaan" (both groups appear to be self-aware, capable of making decisions for them and hence subject to facing the consequences of their deeds). It is interesting to note that the root of the word "Saqalaan" is "Saql" and the meanings are gravity (or gravitational pull). The two groups, the Humans and the Jinn(s) are then asked in verse 32 "which is it, of the favors of your Lord that ye deny?" Finally, there is an explicit address to the two self-aware groups bounded by gravitational pull in verse thirty-three (33) challenging them to proceed with space exploration. "O company of Jinn and Men, if ye have power to penetrate (all) regions of the Heavens and the Earth, then penetrate (them)! Ye will never penetrate them save with (our) sanction". It should be further noted that the verse 33 uses the word "Aqtaar" which is the plural of "qutr" meaning twice the radius, or diameter, of a spheroid in describing the shape of the Earth and the Heavens. So it literally says penetrate "Aqtaar" or go from one side to the other, go through (and around) the Earth and the Heavens. It is a challenge thrown knowing full well that humans are gravity bound – and the glad tiding, that the burden is to be lifted soon. Thus if one was looking to colonize any planet, the verses (LV 31-33) provide ample justification to do so. This will be a good point to repeat the goal. *We would like to reach the "Heaven", (paradise) with our own strength; without being dead, without the arrival of the "day of Judgment" and without waiting for the arrival of Jesus on earth.* If that is the goal, why look for any religious justification? The answer is that perhaps we have understood the religion in the context of the Planet Earth and are stuck in the past. While some of the inertia may be good and that change should be well thought out and understood, absolute inertia is neither good nor justifiable based upon PCI.

Please note that we will not dwell on the religious concept of seven heavens and the like number of "earths". They may have spiritual meanings but to me they are a connection with the history (or just a convention) just as the seven days in a week are regardless of the number of "periods" it took the Almighty to create the Heavens and the Earth. However, let us discuss the "physics" of paradise very briefly. We know that our Universe is expanding and the Quran affirms it; (LI-47) "We have built the Universe with might and We it is who make the vast extent (thereof expanding)". So our religious horizon and paradise must also expand. Further in (III 133) "And vie one with another for forgiveness from your Lord, *and for a Paradise as wide as are the heavens and the earth*, prepared for those who ward off (evil)". Hence, in an expanding universe, the paradise must also expand to accommodate the good that we will *physically* bring to the cosmos, which brings us to the next section.

2.2   The Scientific and Secular Perspectives

Clearly we have to justify our conquest and colonization of Mars for reasons other than spiritual as well. The most important reason in my mind is the continuation of scientific work, as PCI requires or we risk perishing.

a) The Exploration has already begun: The reality that an exploration program already exists is nine tenths of the justification that we explore Mars. It will be futile and unwise to stop it. Many spacecraft (using spacecraft as a generic term) have already been manufactured, failed in their mission, and many more have been successful. The first successful "fly-by" mission to Mars was that of "Mariner 4" by NASA in 1964. The next mission yielding significant information from the twin "Orbiters", each having a "Lander" came in 1975. The two Landers touched down in 1976 and remained operational for several years. The Soviet effort of 1988 (called Phobos I and II) was only partially successful but it did manage to photograph the larger moon "Phobos". The most recent mission to Mars was the NASA Phoenix Mars Lander, which launched August 4, 2007 and arrived on the north polar region of Mars on May 25, 2008. Work continues not only to land on Mars but to also find and explore other wonders in the solar system close to Mars. For example the "Dawn" spacecraft flew near the

Planet Mars on its way to the two most massive asteroids in the main belt called "VESTA" and "CERES" in early 2009. The Ceres is so big it was thought to be a planet for a little while. There are several non-human future missions to Mars that have been scheduled already and this is good news. But it is the manned space flight to Mars that people like me will not see in their lifetime as it is insanely more important to waste trillion plus dollars per decade to fight silly wars on the planet Earth. The humanity will ultimately learn its flaws, we still hope. President Obama has recently added "speculation" that the goal is to reach Mars in mid-2030 and promised to add $6 billion of pittance over next several years towards research.

b) Sustain human life: If a catastrophic accident happens on the planet Earth, the chances of human race becoming extinct are significant. A small colony on Mars reduces those chances of human extinction. It allows for the time to *heal* the Earth. Such accidents are of three possible types. One, a very large asteroid hits the earth despite our efforts to "destroy" or divert it which results in massive upheaval causing earthquakes and a chain reaction of Earth's tectonic plates become uncontrollable. Two, an insane nuclear war of thousands of potent arsenal breaks out and wipes out civilization, makes the Earth barren for a hundred or so years. Three, a germ warfare breaks out and wipes most of humanity by design or by accident. We all know that our presence and history on the planet is very short; we are very good at shedding blood, and quite capable of self-destruction. Even a very small colony on Mars gives us hope that we will reduce the chances of an internal conflict as the nations focus on potential benefits to all humanity and the internal political bickering subsides. Also, on a very long term basis, as the Sun begins to expand (into a Red Giant) it will swallow Mercury and the planet Venus while the planet Earth's fate is unclear; and then the habitable zone would eventually move farther out to Mars.

c) Advancing Scientific Knowledge: While advancing scientific knowledge is common sense there are significant scientific and engineering challenges upon us to overcome as a result of the exploration to Mars. New scientific theories in biology and botany are likely, in trying to make the planet habitable. Medical science has a huge potential and various experiments concerning evolution can be worked upon. In this section one such experiment is suggested: Establishing a new Calendar based upon the unperturbed cycle of the human biological clock (as suggested in reference 1above). Some of this will be pioneering work and can be documented since the clock cycle is not exactly twenty four hours. Also note that the day on Mars is longer by about thirty-nine minutes. Therefore it is good time to set calendar based upon the *unperturbed* (or unadjusted) cycle of the human clock. A specific chapter in this book is devoted to the measurement of time and the new calendar on Mars. Another experiment that has validity has to do with the determination of the optimum buildable depth to increase atmospheric pressure and the value of "g" (the equivalent of 9.8 m/s^2) on the surface of the Earth. The idea is to "feel" as close to being on the Earth as is possible in our daily activities. Please note that in the beginning pressurized "domes" will have to be constructed as the atmospheric pressure on Mars cannot be tolerated without special suits. Another area for exploration is geological. We know certain features such as *the mountains* on the surface i.e. Olympus Mons the highest known mountain in the Solar System, and Vallis Marineris, the largest canyon. Similarly the smooth Borealis Basin in the northern hemisphere of the planet may be a giant impact feature. What is hidden just below the surface of the planet is of immense interest to the scientists. Other scientists and fiction writers have suggested a process called "Terraforming" to make the planet Mars habitable. The scientists and engineers will have their hands full in any such process overcoming shortages of water, oxygen, magnetic shield, and many other obstacles in any actions anticipated in Martian Terraforming.

d) Interplanetary trade: The concept of global trade has been very successful recently and given rise to new alliances and realities. The Europeans may have realized it in depth and are among the first to begin to surrender their "country" sovereignty in favor of a common union and currency. In the United States, the interstate commerce has withstood the test of time and more recently, despite its flaws, has continued to honor the North American Free Trade Alliance. One of the potentials of a Martian colony is to engage in trade with Earth. Since the Earthlings will have realized the importance of trade (by using a common currency etc. for a better future, we hope), the trade with Mars will be extremely beneficial to humanity. There are energy production issues where a possibility of using Deuterium (D or ^2H) on Mars is a possibility in a controlled fusion reactor. Similarly precious metals such as gold, platinum, and others have also been mentioned. It should be noted that bringing things back from Mars is cheaper than taking things to Mars due to lift-off gravitational escape velocity (5 km/s for Mars Versus 11 km/s for Earth) and other issues. One of the main advantages of the resulting technology development (and the trucking abilities of the retiring Space Shuttle is a good example) is to get closer to the main asteroid belt between Mars and Jupiter. There are asteroids that may be mined and highly concentrated iron and other metals readily brought to Mars, processed and then shipped to Earth. This is essential as the resources on Earth are dwindling fast.

e) No other candidate in the Solar System: We can debate all we want but the fact remains that there is no other planet more suitable. We wish that Venus was more hospitable but it is not. Earth's Moon does not have enough gravity and we will not be sure footed. It has no atmosphere and many other shortcomings. Yes it is the closest thing to us and may be a short hop but not much else. Mars appears to have significant advantages over Moon and relatively speaking the most suitable candidate for human colonization. Relative to the Earth there are many disadvantages that must be overcome. The list is very long; a few are mentioned here to remove nostalgia. For starters, the atmospheric pressure is only 6 mbar (cannot survive without pressure suits) compared to Earth's surface of 1013 mbar. There is no oxygen to speak of in the atmosphere. It is extremely cold, and there is no Florida to run to. Even the Arabia Terra is very cold (around minus -69⁰C) and needs an underground pressurized shelter. So while Mars has many advantages, there is significant work and research required to set up stations to begin the long and daunting process of making the planet habitable on a larger scale

f) Other Planets in other Star Systems: It is only a matter of time (and possibly within my lifetime) that an Earth like planet will be found in a Solar System like arrangement. Assuming it is not too far away (i.e. 50 light years or less); it will be natural curiosity to visit. Please note that the nearest neighboring star(s) is the Alpha Centauri system about four light years away. Admittedly, the technology does not exist at the moment, but it may be possible to safely speed up to half the speed of light, and thus be able to visit such a location in an expanding life span of the human species. For example Michael Mayor of Geneva University has reported a planet in the nearby star called Gliese 581. The planet is about seven times the mass of the Earth and orbits at the right distance for liquid water. If this is true, the planet may actually be supporting life as we speak. Now the discovery of certain planets in the star system called Corot-7 some 500 light years away is good to know, but even robotic exploration is distant. Recently, NASA has sent a satellite called "Kepler" with a good telescope to look for "transiting planets". It is slated to scan some 100,000 stars. So the discovery phase of earth like planets is in full swing. If exploration is our destiny, let us begin with our neighbor where we shall feel a little lighter (less weight), a little nimble, and learn to crawl, before walking and running.

g) Artistic Reasons and to tell the truth: This is the last but not the least of the imposing reasons why we should explore and colonize the Planet Mars. I am sure it will give us a new frontier to become and produce a new class of artisan, a new class of painter, a new class of laborer; a new and a more productive member of a just society. It will allow us to ponder over the destiny and further open the secrets of life yet hidden, and take us to a new plateau and new heights. It will produce its own literary giants, new Michelangelo's and new Al-Khwarizmi's[5]. When humans suffer sufficiently and the calamity is severe, among the first to arrive on the scene, besides emergency workers, are the people representing the medium of communications. This includes both official and unofficial press and others. The idea is to visually, verbally, and in writing present the truth to humanity at large (even though sometimes the truth hurts and is suppressed). It should be further noted that normal communication when the Sun is in the way and the planet is at its farthest, it could take twenty-two minutes just to say hello to your child on the planet Earth. Then there is a small period when no communications is possible with Mars. That should be resolved first. Let us bring ourselves one step closer to the paradise[6] promised by colonizing Mars, and the probability of human extinction will forever be one-half of whatever it is today.

2.3   Concluding Chapter II

The ethical thing to do is to continue to move forward with the exploration of the universe in general and Mars in particular. This is after all the spiritual and physical data has been looked at to the best of my abilities. One note of caution must be penned. The fact that we will shed blood wherever we go is well known. So what can we do to avoid it and create laws (with teeth) before we land on Mars and solemnly practice them after landing. It is true that it is highly unlikely that humanity will shed blood as soon as it arrives on Mars, we cannot take chances. Can we come up with

---

5   The west knows about Michelangelo as one of the greatest artists of all times. Al-Khwarizmi is just as well known in the East as having invented Algebra.
6   Paradise in my mind is also equal to guaranteed, forever survival and self-sufficiency of humanity.

a number of suggestions? Constant recitations and reiterations of the commandments will not guarantee anything, promulgation of death penalty does not guarantee it, and nothing we can think of, that will prevent this insanity for sure. *My only suggestion will be to watch for signs of insanity, forgetfulness (as in Alzheimer disease etc.) and do not wait for things to happen. The wilderness plays tricks on our mind, and we cannot take anything for granted. The new planet is completely barren and it is wilder than anything in our knowledge on Earth. This one is for the psychiatrists to ponder on, theologians to think through and anyone else who cares.* On Earth, on an individual level, we take no preventative actions even when it is clear (it becomes clear usually as a result of hindsight) that blood will be shed. We kill for the sport. We enshrine in our constitution to be able to carry guns. On an international basis, we are in a hurry to start a war and take forever to stop it; we fight civil wars, killing on a mass scale that takes precedence. We also know that folks in Hollywood have made billions depicting wars in outer space, and we have enjoyed watching them. So it is certain that wars will be fought in outer space. Thus it is our duty to establish methods of prevention of individual and mass killings. As an example we should ban the use of atomic arsenal in outer space, which is designed only for mass killings and collective punishment. Law and order and courts are not enough – a new mechanism of crime prevention must be thought of; think of PCI. Finally, I am excited and so should be the rest of humanity that we have a good chance to get to Mars and should "soon" start setting up a new just civilization free from coercion and hatred.

Accelerating to light speed yet still a dream
We continue to pull our hair, sometimes even scream
But giving up isn't in our nature, man is on the move
Been through the ringer before, understanding every groove

Migration beyond the Earth actively being planned
Methods being worked together processes being canned
Water is our lifeline and it isn't available there much
It is extracted and manufactured but isn't found as such

Setting up of these outposts, difficult it is to pull
Nations are joined together, cooperation is full
Observation posts are orbiting Martian Space
Designs out of this world with beauty and Grace

You can see that I am now getting carried away in nostalgia thinking about the time when the universe did not exist. Then it was a mere dense point source at Big Bang and when there were no signs of biological life; only building blocks. Then someone "created" the universe with the big bang in "six periods" and throughout the history and even now continues to expand it physically and also our thought horizons. The creation of our solar system clearly (or of it naturally coming together) is much later than that initial start at Big Bang. The known edge of the universe is some fourteen billions light years away. We also know that our solar system is no more than five billion years old.

I must share with you this tradition, while I am in this nostalgia or somewhat mystical state. There is a famous tradition of Mohammad's encounter with angel Gabriel. They, this author thinks, are talking about the age of the Universe (or may be our Solar System), and we paraphrase; Mohammad asks Gabriel "How old are you"? Gabriel responds, "Well! I don't know my age, but I can tell you how old *I* am this way. ***You see O Prophet of God - the place where I live in the cosmos, there is a star that comes on the horizon once in 70,000 Earth years; and I have seen this star rise 70,000 times with my own eyes".*** So, if we assume that the reason for the story was to talk about the star as being relevant to the human evolution, and indeed they were talking about the age of *our star, our Sun.* Then, by this calculation, the Sun is almost 4.9 Billion Earth years old (scientists now estimate the age of our sun to be 4.6 billion Earth years, and who knows they might revise it again but the number is remarkably close, to be told as a layman tradition). There are other fascinating stories from all these holy books of all the religions; suffice it to say: It has taken five billion years for us to lie in the dust, as atoms, as molecules, and proteins, as bacteria from the time the Sun took its final shape, the Earth and the Moon formed and for us to have evolved. *Finally,* *we hope and pray that the Almighty will allow us, the biological life, to have a positive impact on the universe, far beyond the solar system, and Mars shall be the first step. Even if there is no biological life out there, which we doubt, it is our duty to introduce life elsewhere.* ***Whether we were "dropped" here, or evolved on the planet Earth, it is part of our mission to take life and plant it wherever it will take hold.***

CHAPTER III

Measuring Time and Establishing a Calendar

3.1   Introduction

It is essential to realize that the time measurement on Mars (or for that matter elsewhere) needs to be revised in favor of man. Currently the time measurement and calendars are based upon our star, the Sun. We must focus on creating a new Calendar that is truly "human" and not derived or based upon heavenly bodies such as the Sun or the Moon – nor it is based upon any specific religion or theology or a sect within a religion. *This Calendar is designed to allow the human evolution to occur at a natural pace, which currently and in all likelihood, is being impeded due to our methodology of measuring the Time. The Calendar is also eternal in the sense that it is applicable at all locations within and outside the Solar System.*

The purpose of measuring time is to allow *human beings* to live their lives in an orderly fashion. It is presumed that they should be getting up at a certain hour, reaching schools, arriving at their places of work, accelerating in a space ship to elongate life, or simply resuming life after any event. It is therefore *obvious* that a *clock should revolve around us, the human beings, and not anything else – not any heavenly body, nor any other device.* Today our clocks are continuously adjusted beyond the Uncertainty Principle. In short, the measurement of time has not been handled as a continuous scientific process - in which, other than uncertainties embedded in the measuring devices, or the principle itself, *intercalation* has been allowed. Further, it is my hypothesis that the measurement of time is actually in the way of normal human evolution i.e. *we are artificially slowing down the pace of human evolution due to the methodology of measurement of time including the practice of intercalation.*

3.2   Martian Calendars

Calendars of today, on Earth, are all revolving around the measurement of Time and its predictability tied to physical phenomenon such as rotation around its axis, rotation around the Sun, aging among the human beings, and so many other things around us. These calendars have always been based upon heavenly bodies i.e. the Sun, and the Moon. As is well known, we are heavily dependent upon the Sun, our largest energy source, and primarily the source of agricultural production. The concept of the "day" is embedded in the revolution of the Earth around its axis. Similarly, the concept of a month is embedded in the Earth's Satellite, the Moon, and its orbit of approximately 29.5 days around it. The year is a bit more complicated in that the Earth revolves around the Sun giving rise to a different value of the year (as opposed to just adding 12 Lunar months) and this happens every year. That is why the leap year is used to adjust the year once every four years. Many, such as Jews and Muslims, use the Lunar-Calendar while others such as Christians use a modified and marginally more accepted form of Solar-Calendar. The latter has been the de-facto commercial calendar for Earth for a time now. The fundamental problem is that the objective of *all* these calendars was never purposely to serve the human body. These calendars are not designed with the *"human machine"* in mind. Today the Atomic Clock measures time much more accurately than anything in the past, it is still just a measuring device for twenty four hours and inconsequential to how the year and its component months are derived. Citizens of Mars cannot, and must not, use the calendar(s) of the Earth. We must examine this issue in a scientific and ethical way.

3.3   The Biological Clock

A Biological Clock is something embedded in the human beings themselves. The building blocks of our body have in them a sense of "time". It is this sense of "time" which forces us to go to sleep, makes us hungry and so on – and if we purposely refuse to abide by these timed signals, decay and destruction of the body hastens. Human evolution on Earth has made the "biological clock" and its cycle *similar* to the clock on which our current calendars are based. *However, the two are not the same.* Also, direct and additive time measurements do not form the basis of *human friendly* calendars and are significantly off the mark both on and certainly beyond the planet Earth. The postulate is that the humans are less efficient as they try to live their life in accordance with the existing calendars, rising and forcefully sleeping with the rise and setting of the Sun. This phenomenon shall become more pronounced in future especially in *outer space* where our daily routines are not necessarily revolving around the heavenly bodies we have grown accustomed to (for example no agriculture to worry about based upon the Sun). This will also become more pronounced as we tinker with the genetics and make changes allowing for a *more rapid and possibly fruitful evolution of humans not only on this planet but elsewhere*. Before we discuss the Biological Clock and resulting Calendar itself we would like to further explain ourselves.

3.4   Principle of Continuous Improvement

The Principle of Continuous Improvement states that all human and non-human creations, arrangements, and inventions will either perish, rendered useless, or improve in form and function. This is directly measurable by choosing a large number of human inventions and following their life cycle. All human inventions and physical entities have life cycles. Example: As fundamental as the number system itself, Roman Numerals are still used in clocks and so on but not in Algebra. This is because the invention of variables and zero led in Algebra to the use of Arabic Numerals. They, the Roman Numerals, only survive as a subset. Further, there is no guarantee that the decimal system shall not be replaced by a different numbering system if the biological structure of humans (as a result of inventive genetic alterations) has changed significantly. It will certainly be very expensive and time consuming to change more axiomatic things, the Principle of Continuous Improvement shall still hold. Non-human physical phenomenon clearly follows the principle and no examples are necessary.

The importance of this principle has to do with our religious and technological formulations. To the technologists and physical scientists this may already be acceptable. It is not so clear in concepts that have *associations* with the Almighty. For the religious student, therefore, it may be postulated that the only real constant is the Almighty, and the Principle of Continuous Improvement applies only to those not directly measuring the Almighty Himself in any form or function. Hopefully we have not yet stepped on contradictions (as God is not measurable – He is infinitely infinite). The question therefore arises which of these *associations* is "He" and which can be regarded as human intellect and or physical. The experimental test can only be performed on a particular physical entity or a human invention over its lifecycle. Certainly, the current forms of Calendar(s), any type of Calendars, are not "Him". Rather they are designed for improvement in form and function, and their variations speak for themselves. The Almighty is independent of time and by definition not measurable. This might seem commonsense but needs to be explicitly stated as religions have aided in the measurements of time before us. We are thus within our right(s) to apply this Principle of Continuous Improvement for improving our Calendars.

3.5   The Earth versus Mars and the Experimental Challenges

The baggage on the Earth in terms of changing the calendar is too large and political will does not exist. The fact is that in the United States (due to economic costs) we have still not changed to metric system and cling to the obsolete foot pound system. Therefore to change to a biological calendar is too costly and will meet every possible resistant. However, on Mars there are no lobbyists and there are no economic costs in not following the clock based upon the Sun. First, the day on Mars is longer than that on the Earth and its year is almost twice as long. It simply does not make sense to cling to the Earth's time-tables. Primarily there are three things we are postulating. First, without the "Biological Calendar" the *Humans are evolving at a slower rate than if we were to follow the biological calendar.* Secondly, human activity (not necessarily a specific individual) is perpetual - and the challenge is to prove that the efficiency of the *human engine* comes closest to *one* only if a biological calendar is followed. Thirdly a human being is not *just* a social group element. Rather, this magnificent machine must *also* work entirely in its own right and should be able to engineer an individual calendar. This last challenge is the most revolutionary on the part of the individual for the benefit of evolution and to simultaneously maintain group and team activity. This means that in all three cases, evolutionary changes may be measurable within individual or in human groups. Please note that we are not proposing a calendar suited to individuals, *at least not yet*. Initially the calendar must be fixed for all based upon the **"Unperturbed Cycle"** of the Biological Clock". At some future point, calendars could (?) be individualized. *I suggest that the new day be set at 24 hours and 15 minutes on Mars rather than following the "Mars's natural day of 24.6 hours". I also humbly suggest that this value be set for extra Earth missions including the International Space Station and elsewhere.*

3.6   The Biological Calendar

Let us first present the salient features of the new Calendar.

1 New Day = ND = One Complete Cycle of the *unperturbed* "Human Biological Clock" measured to the tick of Uncertainty Principle.

1 New Hour = NH = One-tenth of ND

1 New Minute = NM = one-hundredth of an NH

1 New Second = NS = one-hundredth of an NM

The NS shall be measured to the nth decimal place, most accurate, by the Atomic Clock (or a future clock) until something like the Uncertainty Principle limits the accuracy.

The New Monthly Period (currently called a Month) = MP = (28 x ND)

The New Human Year = NY = (12 x MP)

I am leaving it to the Experimental Bio-Physicist to determine the exact "value" in terms of time quantum and hence ND on Mars – thus defining "One Complete Cycle" of the *unperturbed* Biological Clock (see below). It should be noted that the word *"unperturbed"* is important as often this "biological clock" is *reset* by external factors such as bright light and so on. The Calendar is **not** different for each individual (that may be more a social question, the next stage, if we are to follow the Principle of Continuous Improvement – hundreds of years from now, an individualized calendar?

Please note that the ND will revolve around a "Solar Day", as we will adjust to the rhythm of the human body – in much the same way as a "Lunar Year" revolves around a "Solar Year" although in the opposite direction on the Earth (but perhaps not on Mars). The concept of ND is the most important new measurement as it configures the Calendar to the cycle of the human beings themselves. It is independent of the specific Star or any other heavenly body. The division of ND in sub-units of 10 and 100 is simply for convenience. It should be noted that for humans the only real measure is the ND. *It is equal to the unperturbed single complete cycle of the "Human Rhythm" measuring the start and end of a human biological day. The new Calendar is simply adding one ND to the next, never adding or subtracting any other value of time once this period is fixed.*

The fixation of MP (28 x ND) is simply to construct a month out of traditional seven days in a week and exactly four weeks in the MP – also an exact number (28 x ND) in one MP and in every MP. The New Year (NY) is simply a larger measure for counting a human being's lifecycle (12x28xND).

Now if someone wants to construct a more "metric" MP (as opposed to 28xND) or NY (12xMP), the possibility exits. Remember what is really being measured is the ND. Time will, forever, be measured in ND, regardless of the motion of any celestial object. Here, I am simply attempting to preserve the current notion of a week, a month, and a year. This is my way of preserving the link with "History of Time". I also believe that the change being proposed here is so radical that preserving a historical link will ease the pain embodied in this drastic change. There are other reasons as well but those are mostly theological and not quite obvious. For example, the Universe was created in six days and the seventh day for rest (Bible) *or six periods as in the Quran. The Quran adopts the seventh day as a link to history but rejects the notion that The Almighty is fatigued and needs rest. But the Quran goes further and forbids intercalation. It also states that there ought to be precisely 12 months in a year* and Christian/Jews agree as well.

3.7   Experiment one

How big is ND? This is yet to be determined in a careful experiment, which may last several decades. Previous experiments may have shown that one unperturbed cycle of human clock is larger than our current value of twenty-four hours. Since the calendar change proposed is extremely significant, an enhanced experiment is suggested:

1.   The first step is the proper synthesis of the biological clock and its components. But ultimately we have to measure the *unperturbed* cycles of human clock on Mars, with its day/

night and "weightless" conditions. My suggestion would be to start the process on Earth and select a random sample of one thousand males and one thousand females at age 18 years or older, if possible those likely to visit Mars, when most growth related *possible* perturbations have ceased – all of the individuals as "fit" as possible. The points on the Globe should be close to the two 'poles', the 'equator', and in between at 45 degrees parallel(s). This should be accurate to the current measurable value possible using atomic clock. The experiment shall last not less than thirty solar days on Earth and Mars.

2. Repeat the above experiments at age 45 years for this sample - the idea being that we want to observe issues, if any, of growing old yet fairly active periods of our life. This is also the span when cell growth ends and decay begins? A small sample should be measured on Mars.

3. Determine the maximum differences between the various sets of measurements, if any. If the unperturbed cycle differences in the two age limits are less than a few seconds, we have achieved the goal. This unperturbed value to the nearest "time quantum" is ND. If the aging during this span alters the cycle for more than a few minutes, then we must determine the mean and find a statistical value of the ND.

Please note that given the sample size, any unexpected reduction in sample size should be small (less than 3 %?) and statistically insignificant. A careful experiment might also show other benefits, which shall not be discussed here. The determination of the ND will be the most important historic fact. The unperturbed value may be more than 24 hours due to historical evidence (Refer to the work of Harvard Professor Charles A. Czeisler, PhD, MD). Its value in current terms of hours, minutes, and seconds is extremely important, as it becomes the new "Human Unit" of time. It is, what it is, and needs to be measured to the limits of the Atomic Clock.

3.8 Experiment two

Let us introduce a concept called Evolutionary Level (EL). One way the evolutionary forces contained in the human being can be exploited, and possibly enhanced if certain conditions are met. For example our eyes - if one were to completely remove light from the experience, the eyes become useless and ultimately lose their benefit. On the other hand, if the light available is only from secondary sources, or mildly bright sources (such as working at night shift forever, and sleeping during the time the Sun is up), the *eye may* after several generations begin to see much like a cat at night. If it turns out to be true, this would be a tremendous gift in various explorations on and beyond the Earth. If we take this concept a step further, with some tinkering of the genetics, the eye might begin to see beyond the visible spectrum. We may want to measure this phenomenon in Alaskan's natives, for example after a long winter without the Sun rising every "24 hour long" day. Similarly, the hearing can be enhanced over several generations and deafness wiped completely. The humans might be able to hear in the sound spectrum currently not audible to human ears. Again this would be very useful for survival on other planetary systems. Mars will be an ideal first location. From a religious point of view as well this enhancements in the five senses shall be welcome, as all religions believe in healing. The calendar being proposed may have other far-reaching benefits. It is therefore incumbent upon us to measure EL.

Thus the second experiment may begin with two different samples of males and females starting at age ten or less for several successive generations for the benefit of humanity. What we will be looking for are the following indications based upon either following or *not* following the new "Biological Calendar" both on Earth and Mars. The recommendation is that half the sample (of men and women) can follow the new calendar while the other half follows the usual solar calendar on Earth (and natural day on Mars 24.6 hours). The key factor(s), we are looking for, is that the samples have measured the Evolutionary Level, EL, (it does matter greatly if they are different for the two samples and please note that the measure is not for the individual's brain performance or an IQ in their daily life, although that would be interesting). The design of the EL and the constituent parameters, and their test is left to the Bio-Physicist. *These variables are possibly the physical human characteristics – not any measure of human intellect.* One of the variables in the EL test should be the cycle of the biological clock itself; another might be any measure/ changes in the human "night vision level", as ND revolves around the Solar Day, any noticeable change in the hearing levels, etc. The EL test should be designed to yield a consistent result – what is being measured is extremely important only in the sense that we are looking to detect the size of the change down the road in the *samples* following the "Biological Calendar" versus not following it. The question if mathematically asked shall be, to see if:

$$(d(\text{EL})/dt)_{\text{Biological Calendar}} > (d(\text{EL})/dt)_{\text{Normal Calendar}} \qquad \text{(Equation 3.1)}$$

Further suggestions are as follows:

1. Half the sample follows the new "Biological Calendar". This means that their day shall be equal to ND. Let us assume that by today's time they will get the equivalent of a slightly larger day than the solar day of 24 hours. Theoretically we will not know this until, the experiment one is complete. There is a very small chance that this may be 24hours -Z Minutes. More likely, ND = 24hours +X minutes (again we don't know this – this is to be determined in Experiment one – the number 24+X is for the purposes of illustration). Then depending upon X, we would have revolved ND around the Solar Day (on Earth) in Y solar days where Y = 24 divided by X. If X is one hour, then ND will revolve in 24 days. If X were just 15 minutes (as we have suggested earlier), then ND will revolve around the solar day in 96 solar days. This experiment would require that the sample of people following the "Biological Calendar" would have work schedules based upon ND. The ND shall start regardless of the position of the Sun or the Moon in the sky based upon the determined Biological Clock cycle. In order to keep the rest of the parameters of work/sleep as close to tradition as possible, their work shift(s) should

be one third of ND (this corresponds to 8 hours in 24). The only other requirement is that the shift(s) be consistently applied. All other activity shall be normal – they choose whatever careers in life they would have normally chosen. In the example where ND=24+X solar hours, this first sample of people will always have close to an additional X amount additional time compared to the folks following the solar day. The other half of the sample follows normal Solar Day and they are simply tracked – they too can do whatever in life they would normally do.

2. The experiment has to determine the "Evolutionary Level" of each member of the sample every few years and detect any changes described in the above equation 3.1.

This essentially concludes for the purposes of this thesis, the Hypothesis, the description of the Theory, and suggested Experimentation. I am open to concrete improvements in the scientific methodology applied in the experiments, by the experimental Bio-physicist or anyone else.

One of the questions that might be answered as a result of these experiments is the effect of space curvatures (or that of gravity embedded in general and special relativistic Physics problem of the motions) of the Sun, and its satellites. Is there a correlation with the human biological clock if it is not exactly equal to twenty-four solar hours? If the humans ever permanently settle on other such combinatory systems, Mars included, will that affect the biological clock? Should we include these perturbations in measuring EL? Clearly, the effects of acceleration in space in reaching Mars will also be felt on the Biological Clock, the question also can be asked; will the clock then reset to its original unperturbed value, and can we remain younger by that amount forever?

3.9 Acceptance by the Theologians

It should be noted that we are not forcing any religion to not follow the motion of the Sun or the Moon. It is extremely difficult to change the system one is used to. For example, how would the Muslims, Christians, or Jews shall fast? From dawn (before sunrise) to dusk (after sunset), in the case of religious folks, this will be an issue. We are reluctant to propose additional experiments for lack of our own understanding of the biological clock. One area of great interest shall be to determine the change in body chemistry (or even EL) over the entire *recycle* of the Lunar Year vis-à-vis the New Year (i.e. in roughly n Lunar Years). During the same period, one also knows the precise nanoseconds each day, at points on Earth and Mars of elapsed time. If it is the period that is important then we can precisely replicate the fasting rhythms for all seasons such as the summer, the winter etc. We can also calculate the fasting day variations, including the minimum and the maximum without ever observing the Sun or the Moon. *Please note that even today, we rely on the clock to start and stop fasting, especially on cloudy days and the Polar Regions.* There is plenty of material for a serious student for useful experimentation (please see appendix II). In matters of religions, though, inertia is good. It may be that going slow is alright, as we want to be very careful before changing a practice – there is no hurry.

3.10   New Day Promulgation and Energy Use

It can be said with certainty that, unless we decide to self-destruct, in which case this thesis is irrelevant, the human dependence on energy will continue to increase. Fire was so important to our ancestors, that in many cultures, it was always kept burning and worshipped. It was not long ago that neighbors borrowed a piece of the fire from each other to ignite their little places of fire for cooking and warmth. The technology of energy consumption has continued to evolve but it is still carried from a place of production to places of consumptions. The electrical grids span vast distances beyond the borders of a single country. It is clear to any thoughtful observer, that more energy is to be used and produced on this planet and more energy is needed to explore other planets. It is unlikely that energy sources of the Earth shall be consumed on other planets, other than for rudimentary exploration purposes. However, any serious effort of exploration or utilization of energy will require local production on other planets. Thus it can be said that human leadership will continue to emphasize and diversify serious intensification of energy production that will dwarf the current "megawatt" world production – much the same way as it now dwarfs all the continuous fire production in all of the temples of the world hundreds of years ago. The point of this discussion is that if the ND were to revolve around the solar day, the energy requirements need be fulfilled at many times the current consumption – this calendar will be only an inspirational reason or perhaps the mother of invention? This calendar, it is hoped, shall lead to a revolution of the human genetics technology as energy needs are finally understood. One might say of the new calendar that it presupposes success in abundant energy production. While that may be a foregone conclusion in a world organized along the Principle of Continuous Improvement (PCI), the new calendar is actually much more far reaching.

3.11   Perpetual Work Place

An aspect of the ND is perpetual work by all at all times. The rest and fun is also perpetual. The farmers, if sufficient energy is available, don't have to wait for ideal weather conditions and so on. *The factories and farms could work round the clock.* ***Different parts of the globe or any other planet need not have time zones hindering communications to limit global economies and production cycles. The entire universe can be simultaneously at the same "hour" as there will be no relevance to how far up the sky, the Sun or any other star was shining (or not). There will be no need for a strange date line and no need for a base time called GMT. Clearly this freedom from Sun and the Moon in itself will be another step in the human evolution, both religious and genetic.***

It must be recognized that certain aspects of life need not change. If someone feels that they must go to church when the Sun is at 15 degrees above the horizon on every Sunday (or another angle for

Friday Prayers) – there is nothing in the calendar to forbid them. Remember that roman numerals are still around and in use. People can pray using the new time or the old – God is always there. What may be important is the relative difference based upon the human activity based upon ND. One critical aspect however, is a vastly large source of energy to implement such a perpetual calendar.

A Non-Saturn-like Ring on the surface of Mars: The ultimate challenge to slowly make Mars habitable is to produce "unlimited amounts of energy" from the local sources. As I have suggested a "triad system" of energy production for Earth, the same system can be implemented on Mars with far less complexity. *The suggestion for Earth was to build one 10 miles wide by 24000 miles "long" (around and along the equator) triad of solar panel, wind, and wave panel apparatus, mostly in the ocean (It may be better to build two five mile rings at 30 degree parallel for scientific reasons). The solar panel ring(s) will produce energy to be distributed by a "global grid". The technology is there to do it now in such a way that simultaneously we can harness wind, and ocean wave energy while erecting solar panels (a diagram has been published and copyrighted in my previous book – see Appendix III).* The energy requirements of the future (not the small amount we produce worldwide today) are to take the current production and multiply it by a billion times and more? It has to come from natural sources of energy such as the stars (or planets that produce vast amounts of energy and are essentially uninhabitable or such combinations as are possible to explore. Can a star be harnessed by human(s) much like an atom has been? Are we God's vicegerent? Have we been thinking too small? Clearly this has to be understood and thought through. We are certain that by using the "Principle of Continuous Improvements" we have work to do. The energy production is thus one of the biggest issues of our time and shall remain forever.

A similar and much less complicated arrangement is also proposed for Mars as lack of an ocean on the planetary surface makes it so much easier. Notice that in order to make the planet (any terrestrial planet) habitable, vast amounts of energy is required. We have a rather special case on the Earth, as we have evolved here, and while electricity is essential to our way of life, humans will not perish without it; not so, on Mars.

3.12   Eating Habits and Evolution of Human Diet

We must also discuss change of eating habits as ND revolves around the Sun. In my childhood I had heard from my grandmother (and she lived to be hundred plus) that eighty percent of the diseases, on a longer term basis, are ultimately caused by irregular stomach. We are normally used to breakfast, lunch, and dinner hours. The thesis is that the change from day to day is much more according to the human clock that it will be more natural to follow ND, reduce disease and elongate life? To follow our own clock to go to sleep, have dinner and so on is only natural. In only a short period of time, the system will begin to regulate much more according to the rhythm of the body, and may actually regulate our stomachs. Clearly the experiments suggested earlier will go a long way in confirming (or refuting) the hypothesis. The human diet has to also go through a far reaching change, if we have to reach much longer average healthy cycles of life and explore distant planets and not just Mars. This diet must also allow us to generate incredible amounts of energies without consuming elephant like quantities and turning obese. Tasty and healthy food thanks to a new calendar?

3.13   Home of the Future

We do not want to go off on a tangent here. But the home must be rethought to fit the life style of the human whose ND is revolving around the solar day. This home must be a model of energy independence relative to the normal mode of carrying electricity from far off places. The home of the future may be consuming far more energy relative to the homes of the day. Again the example of the home of a few centuries ago, burning wood for energy production vis-à-vis modern times will go a long way. ND being proposed will force us to take a quantum jump in the technology of the construction

The Ethics of the Colonization of Mars

of our homes, creation of dark and lit areas and many other changes, creation of an environment that mimics the day and night conditions while saving energy.

3.14   In the Beginning of the ND

The economic impact will be negligible on Mars and modest on Earth in the beginning. The experiments suggested earlier will take many decades to complete. However, it will be obvious very quickly if the results are pointing in the right direction. A small worldwide laboratory may be created to start trading at the same hour of the ND and so on. It may be possible to start global communication and clock synchronization based upon some agreed standards allowing for revolution of ND around the solar day, even if the difference between 24 hours and ND is as little as 15 minutes.

3.15   Worldwide Budgetary Cycles

One interesting aspect of such a calendar might be to "agree" upon parameters of a universal budgetary cycle for all the countries of the world. Please note that we are not talking about the corporate specific start and end months of the yearly budget. That freedom may still exist to keep advantages of a distributed versus a centralized corporate economic model. This is to extricate ourselves of the "fiscal" government budgets whose origins may have been productivity cycle of antiquity i.e. agriculture production cycles. The new fiscal year should be based upon the ND. The parameters we are talking about are those based upon productivity monthly data as they are now based always on a fixed value of ND and 28*ND as the New Month. Perpetual production cycles may also streamline fluctuations and a global economy where fundamentals are pegged to a streamlined calendar of all the nations – the production in February will always equal to production in March (in its basis) as the number of days and shifts will be identical - we eliminate an unnecessary concept of the leap year – all monthly and yearly statistical data will be much simplified.

The markets can also ***trade at the same hour on interplanetary scale.*** ***This will save the corporations and the government(s) immense*** ***amounts of money that can be used for research and development*** ***and for exploration of the solar system and beyond.***

3.16   Concluding chapter III

We have proposed a challenge to the humanity. An effort must be made to perform the experiments suggested and funds must be earmarked by one or multiple governments, using their own research organizations. If we just conducted research on the energy source being proposed here, we believe that earth will never run out of energy. The Sun will continue to burn for our benefit for a long time, the oceans of the Earth, and the wind patterns will persist, and so we must rise to the challenge. Both the planets Earth and, in our humble opinion, the planet Mars are in that "goldilocks" zone where life has all the opportunities to thrive and live in peace. The suggestions and experiments work for both the planets. Since any significant change on Earth can be very costly, in the industrial sense (note that in the USA, we have yet to change to Metric System of measurements), the planet Mars might be a good start and no change will be required, just changing our own clocks and the mindset along with it. The exact time of Mars's revolution on its axis as well as around the Sun does not fit our standards on Earth anyway. The time of change will then be achieved with ease. We leave you mired in thoughts.

CHAPTER IV

Basic Data on Mars & Comparison with Earth

4.1   The Sizes

In talking about the size of Mars compared to Earth, there are a number of things to remember. The first is general size comparison of the diameter of the two planets. The Earth is quite a bit bigger than Mars. The diameter at the equator of the Earth is about 7,922 miles (12,756 km) and for Mars it is 4,217 miles (6,786 km). So Mars is about 53% that of the Earth in this sense. But hold on to your horses, this does not mean that the surface areas are similar. In fact, Mars is only about 28% percent that of the Earth. In other words roughly the size of land area of the earth (30%). Interestingly it is also the size of the Pacific Ocean's surface area (30%). This is very fascinating; potentially, we have twice the habitable land between Mars and Earth. Now, on Mars at the moment none of the surface area is covered by water. But if we make it habitable, a small portion (2%?) might be dedicated to water supply purposes, and in fact regarded as water logged.

We would like to take a small tangent at this point to compare ourselves to the cosmos, and how insignificant we are. The volume and the size of a human fertilized egg are insignificant compared to the size of a fully grown man. Clearly men are insignificant in size, compared to our planet. Our planet is insignificant in comparison to the largest planet Jupiter. The planet Jupiter is insignificant relative to our Sun. Our sun is insignificant compared to larger stars such as Arcturus or Antares. Even the largest stars are insignificant relative to the black-hole at the center of our galaxy. Yet the insignificant fertilized egg is about to become one of the most significant entities in the cosmos, if it can figure out how to take the first step, and the next, and make another planet habitable for our species (Chapter V).

4.2   The Distances

Let us begin with the Earth and the Mars on the same side of the Sun, as seen from above, as if all three objects are lined up in a row. This happens once about every 26 months. But the distance from Earth to Mars (in this lineup) is not always the same. On 27 August 2010, Mars was about 34.65 million miles (55.76 million kilometers) away. Back, in 2001, there was a separation of more than 41 million miles (67 million kilometers). In 1995, the distance between the two worlds was nearly double what it was in 2010. The physics is complex but what's behind the seemingly erratic behavior is that a number of things have to line up (with the Sun) at the same time. The measurements involved in the close approaches every 26 months vary because during a year, the Earth's distance from the Sun varies by nearly 2 percent from its average, greater and smaller, and for Mars the distance from the Sun is even worse; it varies from its average by more than 9 percent, greater and smaller. The idea is to get the furthest part of Earth's orbit from the Sun (its aphelion) to line up with the closest part of Mars orbit to the Sun (its perihelion). Thus a journey to Mars will be "fast" under these conditions, i.e. the two are closest, which will not happen every year.

4.3   The Temperature Comparison

The range of temperature on the Earth surface varies from -88 C to +58 degrees centigrade (-127 to 136 degrees Fahrenheit), although there is a claim that a NASA satellite may have recorded in 2005 a temperature reading as high as 71 degrees centigrade in an Iranian (LUT) desert. So in places it is extremely hot. On Mars though, the range of temperatures is even wider, as it is further away from the Sun. The temperature on Mars varies from -125 to 22 degrees centigrade (-194 to 72 degrees Fahrenheit). While humans will not survive long in the cold extremes, it is possible to experience "moderate" temperature regions on Mars depending on how far away (and the orientation) is from the Sun that day. Clearly, the folks on Earth think of 72 degrees Fahrenheit or 22 degrees centigrade as quite normal and indeed pleasant. While some might argue of the cold extremes, and the associated dangers of accidents, vast areas can be made habitable. It is thus possible to dream about living there although a lot needs to be done to make the planet habitable. While we are talking about temperatures on Mars, and they are important, in the beginning at least, even in the moderate temperature zones, people will live in temperature and pressure controlled underground domes. It will take a long and continuous effort before humans will walk freely without pressurized garments on the surface in mild weather and actually enjoy it.

4.4   The Gravity and Atmospheric Pressure Comparisons

Perhaps the most important comparison is that of the feel of our own "weight" on the surface of Mars, and the associated "light-headed-ness". The bad news as written elsewhere is that we cannot walk on the surface without a pressurized space suit. New technologies are being tried for this purpose and promising garments will be there when we are ready. The main problem on the atmospheric pressure is well known; compared to Earth's 1000 millibars on the ocean surface to just under seven millibars on the surface of Mars. So walking without any pressurized garments will be suicidal. Furthermore, the gravity is only about 38% of the Earth. So while three hundred pounders on

Earth will feel to be around 120 pounds only on Mars, all is not "feel good lighter" turning from being obese to a "normal" man. Again experiments are underway to determine the proper weight additions to the pressurized suits to make us feel and work "normally". One has to get used to the idea of living "underground" where the atmospheric and gravitational conditions are more to the liking of someone who has grown up on Earth. It is unclear as to what the effects of being born on Mars and growing up there will mean. Will we slowly evolve and begin to acquire larger mass and hence weights or simply shrink to adjust to lesser demands of body energy consumption? How will the evolution of man on Mars will affect trips to the Earth and so on; these are all interesting questions which we will have to grapple with.

4.5   The Rotation Periods, and the Moons,

Our day on earth is 24 hours long. The Martian day is bit longer, by about 40 minutes, for those Sun tanners. But for some of us it is the Martian Year which is about twice as long as that on the Earth that is such great news. So if you are forty years old on the Earth, in Martian Calendar only about twenty years old. It might just seem we are back into our youth. And don't forget a few milliseconds when our biological clock went slower, as we were accelerating in the spaceship. In the Islamic theology there is conjecture that when we enter paradise, we will be young and youthful. Perhaps, we have moved closer to the paradise as we land on Mars. As many know, there is only one moon orbiting the Earth. But Mars has two relatively speaking small moons. The Earth's moon has seen so much poetry written (eastern languages). The Jews and the Muslims, and many others still measure their time based upon the revolution of the Moon around the Earth. In fact, the concept of the "Month" is due to our Moon. There are twelve revolutions of the Moon around the Earth, in one year. We still use this concept, as the solar month is really not based upon any heavenly body revolutions, and we can adjust these months, leap year, and what have you. Therefore, it remains to be seen as to what happens to our poetry, our music, and our culture, as we begin to colonize Mars. The Mars moons are called Phobos and Deimos and unlike our Moon, are thought to be recently captured asteroids. We will leave those poetically inclined to get details of these moons elsewhere. Just remember that from the surface of the planet, the moon Phobos looks small compared to our Moon and the one called Deimos (so small) looks more like a star than a moon from the surface. We leave you with some comparisons in the tables below. Again this is to start thinking scientifically for all on Earth, and get excited about the possibilities and challenges of living Mars.

General Data (Source: NASA)	Mars	Earth
Mass (10^{24} kg)	0.64185	5.9736
Volume (10^{10} km^3)	16.318	108.321
Equatorial radius (km)	3396.2	6378.1
Polar radius (km)	3376.2	6356.8
Volumetric mean radius (km)	3389.5	6371.0
Core radius (km)	1700	3485
Flattening	0.00589	0.00335
Mean density (kg/m^3)	3933	5515
Surface gravity (m/s^2)	3.71	9.80
Surface acceleration (m/s^2)	3.69	9.78
Escape velocity (km/s)	5.03	11.19

Orbital Data	Mars	Earth
Semi-major axis (10^6 km)	227.92	149.60
Sidereal orbit period (days)	686.980	365.256
Tropical orbit period (days)	686.973	365.242
Perihelion (10^6 km)	206.62	147.09
Aphelion (10^6 km)	249.23	152.10
Mean orbital velocity (km/s)	24.13	29.78
Max orbital velocity (km/s)	26.50	30.29
Min orbital velocity (km/s)	21.97	29.29
Length of day (hours)	24.6597	24.0000
Obliquity to orbit (degrees)	25.19	23.44

Data on the Moons of Mars (Source: NASA)

	Phobos	Deimos
Semi-major axis from the center of Mars (km)	9378	23459
Sidereal orbit period (days)	0.31891	1.26244
Sidereal rotation period (days)	0.31891	1.26244
Orbital inclination (Degrees)	1.08	1.79
Orbital eccentricity	0.0151	0.0005
Major axis radius (km)	13.4	7.5
Median axis radius (km)	11.2	6.1
Minor axis radius (km)	9.2	5.2
Mass (10^{15} kg)	10.6	2.4
Mean density (kg/m³)	1900	1750

4.6 Other Comparisons

There is much more data to be compared. However, we have purposely avoided this so that the book does not become a purely scientific treatise. There are ever increasing details of the surface of the planet Mars. There are continuing missions shedding more and clear light on the similarities and differences between the two planets; in fact an entire book is needed just for that. One concern we have is the lack of a "shield" to safeguard human species. Examples of the components will be that of the magnetic shield and a thick atmosphere. These are essential elements to allow the evolution to proceed without always worrying about its direction. We have discussed this in detail in the next chapter to plan and ultimately erect a "shield" as part of the rehabilitation project of the planet Mars.

Despite all these differences on Mars, there are significant similarities to the Earth's climate. For example, it has polar ice (on both poles). It has observable presence of weather patterns. Mars, like Earth, has recognizable and predictable seasonal and weather changes. It has ice clouds (aphelion approach) and dust storms (perihelion approach) that may last for months (?). Mariner 9 probe in 1971 observed such a storm, as recently did the Hubble telescope. In mid-2007 a planet-wide dust storm posed a serious threat to the solar-powered "Spirit" and "Opportunity" Rovers by reducing the amount of energy provided by the solar panels. Olympus Mons and other volcanoes in the so called "Tharsis" region also affect the weather and climate of the planet considerably.

Despite all the differences and similarities, we, the humans have made up our minds that we are indeed God's vicegerent and a deputy appointed to act on His authority; to go where angels live and where paradise awaits us. We know that one generation cannot accomplish this great goal but we have preserved the base of knowledge and are imparting it to the next generation.

We end this chapter by urging the reader to keep faith and keep pressing the "authorities" to keep pushing the human frontier on to other star systems, spiritually, physically, and peacefully . Let me go out on a limb and say that my grandchildren will actually see a successful mission to Mars, and their grandchildren will see an outline of the Martian Colony, a place of worship, a cemetery and a worthy governmental structure.

CHAPTER V

Making Mars Habitable

5.0 Historical Perspective

After several failed attempts by the former USSR and the United States starting in 1960, the first truly successful mission was conducted by Mariner 4 from the United States - (November 28, 1964- December 20, 1967). According to a NASA press release, Mariner 4 arrived at Mars on July 14, 1965 and passed within 6,118 miles of the planet's surface after an eight month journey. This mission provided the first close-up images of the surface of Mars. The spacecraft returned 22 close-up photos showing a cratered surface. The atmosphere was confirmed to be composed of carbon dioxide with a surface pressure in the range of 5-10 mbar. A small magnetic (intrinsic?) field was detected.

The next significant milestone was achieved, when Mariner 9 arrived at Mars on November 3, 1971. It was then placed into orbit on November 24. This was the first US spacecraft to enter an orbit around a planet other than Earth. A huge dust storm was in progress on the planet upon its arrival. Many of the scientific experiments were delayed until the storm had subsided. The first hi-resolution images of the moons Phobos and Deimos were taken. River and channel like features were discovered. The author has mentioned the storm in one of his publications in an epic poem on mentally disabled, contained in "Swad-e-Hoor" published from Delhi.

Next significant success was achieved by Viking 1 and 2. They consisted of an orbiter and a lander. Viking 1 was launched from the Kennedy Space Center, on August 20, 1975, on the trip to Mars. The craft went into the orbit about the planet on June 19, 1976. The associated lander touched down on July 20, 1976 on what is known as the Golden Plains. Viking 2 was launched for Mars on November 9, 1975. The landers provided detailed color panoramic views of the Martian terrain. They also monitored the Martian weather. The orbiters mapped the Mars's surface, acquiring thousands of images.

Following these successes and many failures by USSR, and the US and even the Japanese, the Mars Pathfinder delivered a stationary lander and a surface rover to the Red Planet on July 4, 1997. The six-wheel rover explored the surface. Next, the Phoenix Mars Lander was launched on August 4, 2007 and landed on Mars on May 25, 2008. It is the first in NASA's Scout Program. Phoenix is studying the habitability potential on Martian Arctic and its icy soil.

Clearly, NASA is committed to the exploration of Mars and we are very thankful to it as well the tax-payer.

5.1   Oxygen and Liquid Water

It is clear that we have a lot of work to do, to make the planet habitable; but when it comes to our survival, we can perform miracles. Moses comes to mind; in the Quran (Chapter II verse 60), the Almighty wants us to "remember that Moses prayed for water for the people: We said "Strike the rock with thy staff", then gushed forth therefrom twelve springs. Each group knew its own place for water". Clearly, life for humans without air and water is not possible. The matter is far more complicated on Mars as the oxygen content in the atmosphere is very little and it is so thin (pressure is low) to the point that man will die without pressurized suits on the surface. Therefore the first job is to make sure that we can land in a safe place on Mars, with little or no environmental hazards. To this end a number of unmanned space missions are required, and much has already been accomplished by NASA and others. These missions will arrange for additional "data gathering" on the surface, and from a safe yet close distance of a few kilometers above the Martian atmosphere.

One other important issue for us to handle is the lack of Nitrogen in the atmosphere of Mars. As you know, Nitrogen is an inert gas, and it controls Oxygen from letting Carbon or any other fuel burn uncontrollably on the surface and in the atmosphere of the Earth. Fortunately there is Carbon Dioxide in the atmosphere. The humans have evolved with just about 20% oxygen in the atmosphere. Imagine if we were living with 100% oxygen all the time, we will oxidize ourselves to death much faster. Thus ultimately the issue is of a careful duplication of the Earth like atmosphere. That is a tall order; and while there is no hard deadline yet, I am afraid that one is coming soon by cosmic standards. If we go by the scriptures we quoted in chapter II, where He promises to lift the burden of Gravity momentarily, by my calculations we have roughly six-hundred years left. In XXXII-5 "He directs the affairs from the Heaves to the Earth: Then it ascends unto Him on a day the measure of which is a thousand years of your reckoning". So in my mind we have roughly six-hundred earth-years to make the planet Mars habitable; plenty of time one might say, probably part of His timetable.

NASA has announced that scans of the Martian geology had revealed the presence of large quantities of water on the planet, paving the way for a manned mission to Mars in the future. But how will these Martian pioneers find the water, and equally importantly, oxygen on the red planet? Scientists Don Sadoway, from MIT in Boston, and Ken Debelak, from Vanderbilt University in Tenessee told a NASA conference on how to do it. Sadoway has designed an electrochemical cell the size of a fridge, which is powered by a small nuclear reactor. Oxide-rich rocks, which make up the surface of Mars, are loaded into the cell which passes a strong current through these rocks, melting them, and releasing oxygen through electrolysis. People need about 3 kilograms of oxygen per day, which the cell should be able to extract from only 8 kilograms of Mars rock. So what about the water? Debelak, the other scientist, has suggested using the same technique employed home here on the Earth to make decaffeinated coffee! By compressing carbon dioxide gas, which makes up most of the atmosphere on Mars, it can be used to dissolve some of the water locked up in minerals and rocks on the planet surface. When the compressed gas has passed over the rock samples it is allowed to expand which releases clean water which can be collected and used.

5.2   Exploration and Vehicle Classification Suggestions

In the meantime, a number of vehicles, that takes advantage of gravity assist need to be designed and manufactured. In fact a whole fleet will have to be built. First a large ISS (International Space Station) class vehicle that goes back and forth between the two planets, **but never lands on either.** At least two such ISS vehicles are required. The wear and tear should be relatively small as they never enter the Earth's atmosphere (or even Mars). *At the risk of showing ignorance, I will convert the current International space station (ISS) to this type of vehicle and build another identical space station very quickly. So one ISS is near Mars and the other near Earth and the two go back and forth but never land on any planet yet both are permanently occupied.* This is perhaps the best way for a person to not be in space too long and make sure that weightlessness and other such health issues can be addressed and we slowly become familiar with the Martian atmosphere and land topography. Next a class of vehicles that connects Earth to ISS; we will call it "Shuttle-E". Shuttle-E merely flies back and forth between ISS and Earth. Also, a fleet of "shuttle-M" is built which are designed to go back and forth between ISS and Mars. Shuttle-E is to have the same cargo capacity as Shuttle-M for obvious reasons but some of the other details can be different. Finally, Transport-a, Transport-b, and Transport-c are designed and manufactured for land transport and other excavation jobs on Mars. We want to add one more vehicle type; we will call it Mega-T (Mega

Transport) to be explained below and will be used to haul Asteroids from the nearby Main Asteroid Belt between Jupiter and Mars.

It is prudent to further state that Shuttle-E, and Shuttle-M should have comparable cargo carrying capacity of C5C (or C5M?) or equivalent Russian aircraft. These vehicles should be able to carry large volume and weight to the ISS and thus bring back cargo coming from the two planets. A bit of a suggestion for the transport-class of vehicles described above; one of the requirements should be, for them to be able to operate when humans are inside, without wearing the pressure suits. Clearly the scheme we have outlined should work. However, I am the first believer in PCI and will look for better or more economical solutions as long as the basic requirements are satisfied. There are numerous complications, but if a Monarch Butterfly can cover two-thousand miles on a wing-span of just five inches, we should be able to do better. Also, in the beginning, no grandiose designs are required; only safety should be the greatest concern. While humanity is taking a chance and is ready to *sacrifice life to save the species*, carelessness cannot be tolerated.

5.3 Building an Atmospheric Shield

As is well known, the Earth is truly a heaven from our point of view. The atmosphere is thick and saves us from many things. The Earth is constantly showered by small meteors and the atmosphere burns most of it down. Also, two thirds of the planet is liquid water and what survives the atmosphere falls harmlessly in the water (possibly adding more to it?). Now, there is atmosphere on Mars but it is thin and most of the 'stuff" falling will not burn and fall on the ground. Any population will then be exposed and in danger of getting wiped out. There is also no deep liquid water and the size of the planet Mars is small compared to Earth. Therefore, one of the issues of habitation is making the atmosphere thick (note that the height of the Martian atmosphere may not be an issue. It is actually higher than that of the Earth 11Km compared to earth's 7 Km). To increase atmospheric pressure is not an easy task, but a long term plan can make it happen. Just like all materials in the physical world, air molecules are affected by the force of gravity. And, as with all objects, the heavier or denser an object is, the greater gravity's pull on it. The idea is simple. Keep adding more Carbon Dioxide to the Martian atmosphere until the atmospheric pressure allows for men to walk without pressurized suits (need not equal that of Earth). Now, the idea may sound absurd, and may be it is, but still worth trying. A side benefit will be to warm up the planet a bit. The oxygen/nitrogen mixture will be ideal, but will mostly escape the Martian atmosphere. CO_2 on the other hand has a molecular weight that is much heavier (44 VS earth's air 28 AMU's) and so likely will not escape Mars altogether. Please note that we still need a breathing mechanism as CO_2 is unfit for that. So then, instead of a bulky pressurized suit, we might wear a very light garment that simply extracts oxygen from the atmosphere, and mixes it with Nitrogen in the garment to provide an ideal breathing mixture. Part of the habitation plan will be to set up (nuclear powered) factories that simply produce a mixture of nitrogen and oxygen to be used for breathing underground where the first colonies will have to be established in various zones for a host of reasons but most importantly for the safety of the crew engaged in the initial exploration and set up. Finally, this heavy addition of CO_2 is like erecting a shield the surface of Mars.

5.4   Underground Living on Mars

Sometimes we learn lessons by accidents and calamities. A relevant case in point is that of the thirty-three Chilean miners who were trapped about two thousand feet below the surface of the Earth, for about seventy days (most of August, September and part of October 2010). This was due to an accident, yet it clearly shows that planned deep underground living is possible; obviously not on day one. There are three reasons why a significant human presence should be planned underground (as opposed to living quarters on the surface), as deep as two to three thousand feet. First, it will be a bit more like the pressure and gravity "feel" of the Earth and properly pressurized. Second, it will protect against an accidental heavy (but small in size) meteor shower/asteroid strike on the surface of Mars, as the asteroid belt is somewhat closer and thirdly, to build pressurized quarters so humans will not have to constantly wear pressurized suits. It is a bit like living in a deep ocean going submarine (but different in the technological sense).

There is another aspect worth mentioning. The underground channels to be built must be designed for humans to escape the planet altogether. In other words, in a natural calamity on Mars, the humanity must be willing and able to escape Mars. Please note that, if the planet Earth is not within reach then we must have an alternate, a close dwarf "planet", in the asteroid belt or even a large body, a huge space station, on a temporary basis. This will be the type of situation, where the planet Earth has also been devastated by a prolonged nuclear war or some other calamity and the human race is about to become extinct.

5.5   Hauling Asteroids from the Main Belt on to Mars

As we know, there is a significant belt of asteroids between Mars and Jupiter. Jupiter is the largest planet in the solar system and about 11 times larger in diameter than the Earth. Mars is the last of the inner planets, and somewhat too small for humans to permanently endure the loss of gravity and so on. A wild idea to explore will be to begin making Mars gain additional mass and hence gravity. If we could design a few Mega-T transporters, we might be able to haul smaller (between 1 and 10 Km diameter asteroids) and gently place them on the surface of Mars, with predetermined locations. For example we might begin on the opposite side of the planet to where Olympus Mons is and so on so as to distribute this mass as uniformly as possible. A Mega-T is a vehicle that looks like an open truck-bed and fitted with rockets. A Mega-T approaches an asteroid and gently envelopes it, in its bed. The transporter then manages the "rock" to assume the orbit of Mars. It then gently places the rock on the surface of Mars. There may be as many as a million candidates to do this. But, if we want to be selective, we can begin hauling "metal-rich" asteroids first. Clearly, a diagnosis of the asteroid's chemical composition is required. Over 200 asteroids are known to be larger than 100 km. While this will be a heroic effort, the increase in Mars's mass will be perhaps just 1-2 percent; every little bit counts. However, the main reason for such a transport design is to gain experience in large scale hauling and ultimately help Mars gain a gravitational pull somewhat more comfortable for human living.

5.6 Water and Seeds from Earth (NOT from Ceres in the Asteroid Belt)

Most scientists may argue with this point, but my view is that Earth's Ocean Water and associated plant life and bacteria should be transported on the planet Mars on a small seedling scale. I am opposed to bringing water from sources other than Earth at least in the beginning. We are not sure of the environment on the moons of Jupiter or Saturn, and much research is required. Also, Ceres is too small and any significant disturbance may be devastating for both Ceres and Mars. The amount of ocean water from the Earth, similarly, should not exceed five to ten million gallons. This will not in any significant way alter the Earth's mass or it's any other properties. However, it will provide (like seeds/water mixture) a laboratory on the Mars to begin or replicating the process of sustaining life in a real sense. Fresh water can be extracted from Mars itself. The location to store the ocean water on Mars needs studies. Clearly, additional analysis is needed to make sure that this simply does not evaporate out of Mars. The lack of any significant magnetic field is a huge problem and must be overcome by a process similar to the description in 5.3 above and very long term planning for increasing gravity (see below also).

5.7 Mars Magnetic Field Generation

It will be very nice to generate an Earth like magnetic field on Mars that will protect us from the solar winds and so on once we inhabit the planet. It will perhaps also deflect the hydrogen molecules from escaping the planet and so on. The theory of Marvin Herndon whether true or not appeals me the most. The challenge is to generate an artificial nuclear powered magnetic shield on Mars, in essence mimicking the Earth. A physical dynamo powered by the reactor(s) shall generate the magnetic field to shield the planet from harmful solar winds. We may have to think more than one such generator(s) and placing them on the surface or underground near the North and South poles of Mars. Clearly, reaching the center of Mars and installing one massive magnetic dynamo will be "impossible".

5.8   Concluding Chapter Five

We conclude this chapter by saying that if we did all that is in this chapter we will have laid the foundation of terra farming on planet Mars. *"In the beginning, there was water, and the water was full of life. The intelligent life was brought from the neighboring planet Earth by the gods. The gods fed all life on the planet Mars, and they liked what they saw. They made a covenant that there will never be a war on this planet. To this day, Mars continues to provide a sanctuary for gods", wrote a historian in the year 5000 from the day of migration.*

CHAPTER VI

Establishing a Corridor of Worship

6.1 The Doctrine of Worship

The basic definition of worship, in my mind, is contained in the act of worship. The act of worship gives me a different perspective on life, something that takes me away from the daily "rat race". During the act of worship, I am *not* busy on a highly competitive basis, slandering others, looking for some scandal, and doing this day after day, *until I have arrived at my grave.* For me worship is to ponder on things that are not part of my daily routine or curriculum. If my worship is part of a curriculum, it is no longer worshipping, it is then some organized stuff that is meant to earn a living or even worse, it is wasting time. Various religions have tried to create a balance between the acts of worship and daily routines, but it seldom works. It is perhaps in our blood, that we want to invent and be knowledgeable, or it is PCI in play; but when we have a problem whose solution is "not possible" we turn to worship. Sometimes, on such occasions, we resort to jugglery, hocus-pocus, or look for miracles. As a result a number of "spiritual" solutions are desirable to console us, make us hopeful after a death, and to resume normal operations after a terrible ordeal. Humans have thus over many thousands of years "invented" every religion known to us today and more are coming as we discover and explore other planets. Therefore we will continue to pray forever. We will always have formal prayer locations and pray in community settings no matter what planet we live on. On Mars, I will pray something like this:

O God! Change my fear of stark loneliness on Mars into love! Have mercy on me and make me among the greatest of the explorers. Make me remember what I have forgotten and give me knowledge of what I have become ignorant of. O God! Make us not the enemy of the other on this planet; make Mars habitable and provide us all the provisions for a pious living as you did for many of us on Earth; raise our conscious on Mars so we don't repeat the mistakes we made on the Planet Earth. O God! Let the soil produce green plants and the meanings of Mars become peace; O Almighty! Let there never be a war on this Planet! Amen!

I will request priests of all the religions to include this in their daily prayers. This will be the common prayer and bind us all who will visit and live on Mars. I am also requesting all who will make decisions to set aside a *Corridor of Worship* on Mars. I know that the current climate is not conducive to any talk of harmony, yet it is a fact that the followers of the three great religions i.e. Islam, Christianity, and Judaism together are a majority of the population on the Earth. It is also a fact that all three originated from the Arabian Peninsula, thus one more dry than the other. *The suggestion therefore is to build this corridor of worship in the Arabian Terra on Mars (see figure III in Appendix III below).*Clearly the precise location shall be determined by a "survey team".

6.2   Building a House of Worship to One God

Perhaps it is wishful thinking on my part that we will come together to build a single house of worship for all religions at this location. But it is worth proposing. This one house will be equal to Jerusalem's temple, Ka'aba, and Saint Peter's Basilica combined and put together on the planet Mars. This will be one house of worship where they will all pray together. Each may have their own timings and hopefully those timings will not be based upon the Sun, or the Martian moon(s) as the time shall be measured on the basis of the 'unperturbed" cycle of the biological clock, discussed elsewhere in this treatise. In order for us to connect with history, the new calendar still has the same concept of seven days i.e. Sunday through Saturday. The major congregations, some day in remote future, shall still be possible without much interference of the other two (Friday, Saturday, and Sunday or maybe there will be just one religion). *The structure built on the surface will be symbolic and skeletal in nature and the real prayers will most likely be hundreds of feet under-ground below the structure in a pressurized environment, at least in the beginning (please refer to chapter V on making Mars Habitable).*

**When alone pray in any direction
Together pray in the same direction**

The Floor Plan of the Worship Corridor – the scale is left to the Architect – recommended 10 kilometers square. The Symbols in their entirety should be visible from a space craft, and landing shuttle runways should be constructed along the Eastern Edge for pilgrims to visit easily (see section 5.2 on vehicle classifications).

Note that the last circle symbolically places the planet Mars around the Sun. Although the real orbits are elliptical, a circle is placed for historical and mathematical simplicity.

6.3   Proposal to Build a "New Ka'aba"

In this section, what is being proposed may be disturbing to some conservative Muslims? But I have a very simple justification. The first house of worship on Mars will be indeed the Ka'aba for Mars. In order to pray individually, there is no need to face the same direction, as God is in any direction. However, in a collective act of worship, a requirement exists to face in the same direction. Since the Earth may not be visible from Mars, at all times, and further it will be somewhere in the Martian Sky, the only solution left is to build a new house of worship. Then from anywhere else on the surface of Mars, one could lead prayers in the direction of this First House of Worship. Muslim scholars can weigh in on this proposal, but we doubt, they have a better solution. This proposal makes both scientific and religious sense. Now, it need not look exactly like the structure on the Earth, my personal preference will be to do just that. Remember, the surface of the planet Mars is extremely cold, and requires a pressurized suit to work or even pray. That is why the structure being proposed is only skeletal, as no roof is required. People will actually pray somewhere in an underground chamber, most likely underneath the skeletal structure, as it will be impossible to pray on the surface without pressurized suits. Not only will these suits be uncomfortable, cleanliness may not be achievable, except for a relatively short period of time. Most of the activity of any kind shall be in special underground structures for a number of reasons (at least in the early days of colonization). Remember that the atmosphere of Mars is very thin and the "shields" that we are used to on Earth, are not there. Thus, collision of a very small outer object (such as even a tiny meteor) may destroy any structure on the surface of the planet. So the object of the house of worship is to not actually pray in large numbers, but establish a symbol of "Oneness of God". While I personally will welcome Jews, and Christians to pray in there or anyone else, or any other place, they are certainly welcome to build a Basilica or the Solomon Temple in the worship corridor.

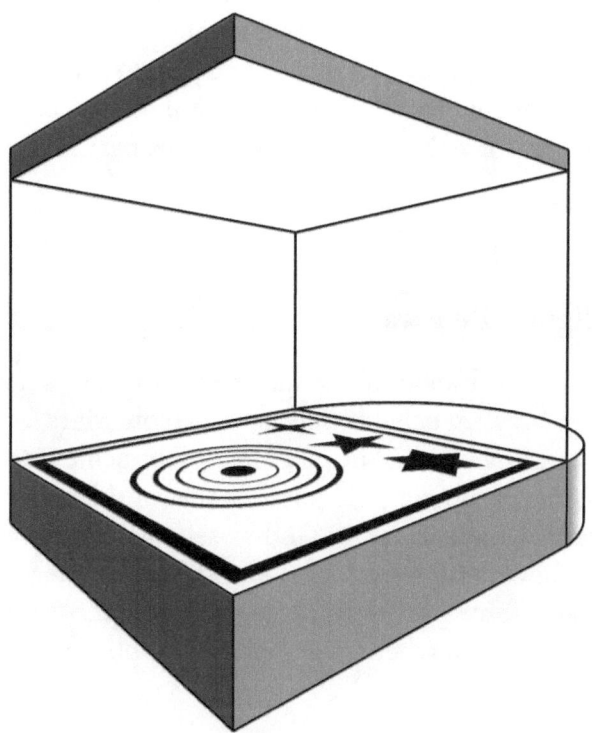

A sturdy "Wire" Frame Model – No roof or walls – The floor has the Martian Symbol; the structure(s) will be sufficiently large to be visible from the space.

Clearly there are people who do not believe that we as a species will live on Mars in any large numbers. They are entitled to their opinion. There are Muslims who do not believe that we will ever pray on another planet. They too are entitled to their beliefs and opinions. The scientists are convinced that the humanity is destined to colonize far beyond the Earth, and that Mars is just the first step. It is thus our duty to point out that not only humans are headed to Mars; we should recognize that we are headed there with our religions (hopefully enlightenment), our heroes, our history, our culture, and certainly our music. I repeat the last sentence in the Martian prayer (section 6.1 above) O Almighty! *Let there never be a war on this Planet! Amen!*

We are therefore convinced that we must build houses of worship, and I am appealing to Saudi establishment, the Pope, and the high priests in Israel to support this section and plan to contribute to build a corridor of worship, that will house not only the places of worship but also the first formal burial grounds in this part of the planetary coordinates.

6.4 Religious Renewal

Some of what we are about to say may be regarded as a deviation by various sects. My firm belief is that as we explore Mars, a significant new understanding of (all?) the religions will be achieved. Central to this is the concept of God (and possibly that of the hereafter). If we were to think of the concept of God as self-sustaining, within and beyond the known universe(s), exploration beyond the planet Earth will most likely force us to rethink such a concept. At a minimum, we will ask ourselves whether or not PCI applies to our *detailed definition* of God.

If it turns out that we begin to acknowledge that we have little understanding, of the *concept* of God, as we acknowledge the relatively unused (or misused) portions of our brain. If it turns out that we begin to acknowledge that our definition of God has been deeply flawed, a religious revival and renewal of a *singular human religion* will move us forward in the right direction. It might propel us to come up with a new paradigm and "structure" that will shed light on the conceptual religious issues that come to us today but are afraid to ask let alone conduct research.

It might tell us when 'inertia" in religious matters is good and when a change is not only essential but the speed with which such a change must be implemented. *If we begin to understand the fundamentals of the religions better; that will give us religious fervor in further exploration of the universe.*

In the second last paragraph of 2.1 above we have quoted from the verses (LV 31-33) in which we (the humans and jinn both explicitly stated to be gravity bound) have been challenged to explore the universe. It is interesting to now talk about LV-35 (please note that verse 34 is identical to verse 32 for emphasizing the obvious). The reason to talk about verse 35 and so on is to show that historical explanations of the literal translations are tilted towards a sinful condition before and after death. Literally, the translation says "There will be sent, against you heat of fire and flash of brass and ye will not escape". In fact verses 37, 39, 41 43, 44 continue what has been described by the historical explanations, as consequence of *sinful disobediences by both the humans and Jinn,* unacceptable to the author. Our view is that humans after being challenged to explore the cosmos are being warned that the usual shield(s), i.e. magnetic, atmospheric, specific gravitational values, and so on that are protecting the planet Earth are not there as soon as you are beyond the usual atmospheric limits. We are being warned that harmful cosmic rays, solar flashes, and so on are extremely harmful. Unless you know what you are doing, you are doomed. Verse 39 warns of the splitting of the "sky". In other words, if we have thought about all these complications before challenging the gravitational and other complexities, then we will be successful. Not only a victory is near (LV-31) but all the benefits that result from such a crucial victory are also for us. That includes a life of luxury and bliss, far beyond what we can imagine, much like the concept of paradise.

The end of this chapter is to digress in just two lines. With exploration of the cosmos, we can be assured that a new *singular human religion* is around the corner, and the first incarnation of that begins at Mars - O Almighty! *Let there never be a war on this Planet! Amen!*

CHAPTER VII

Preservation of the Languages and the Historical Context

An extremely important aspect of the human civilization is the ability to speak, read, write, and to standardize the letters and words in a language, and to continually enhance the vocabulary; and in so doing to further the evolution through cross breeding of languages. Humanity has relentlessly made changes in the meanings of (sometimes) the same words and idioms. In 1611 A.D., when King James Version was being printed the first time, the words "to suffer" meant "to permit or to allow". Today it means to be in misery. Similarly, the words become obsolete yet they remain preserved; Thee, Thou, Thy, Thine, or Trow (meaning to think) all belong to the English Language history. In this sense, the Bible has played a great role in the preservation and enhancement of the English language. Clearly, the effort to "humanize" the bible started with John Wycliffe long before the year 1611. Today the translation of the Bible (or for that matter the Quran) in various languages is considered adding to the promulgation of the religious understanding beyond just the sectarian interpretations. It also leads to new research in the old language and its connection to the modernity.

Midrash (plural Midrashim) in Hebrew means "to investigate" or "to study" or broadly speaking "to seek knowledge", I have read somewhere. Similarly the word "madrasah" in Arabic means a place of learning; how close in meanings, an example of cross breeding of both the humans and their languages. And so there are Semitic and Indo-Germanic languages and Asian languages and so on. In a uniform cultural environment we barely use a thousand words to communicate with each other. The human languages are so rich that if we were to add all the words of all the languages, they will be several million. The English language alone approaches a million words, assuming scientific vocabulary is included. We have come to a point that no one person shall use all the words in the English language and a full understanding of all the words and their meaning will require more than a life time. Yet the English language (and others) remains a fertile ground to add new vocabulary as the humans arrive on Mars and find strange things.

There is no question that on Mars, "new" languages will slowly take shape. They will have different words; those will be an addition to the human civilization. There will be new nouns, pronouns, new verbs and so on. The idea is to carefully preserve the current and the new meanings for posterity much the same way that we built a library of congress. A new "Martian Library of Revelations" will have to be constructed where all the scriptures, their original and translated texts, presumed origins of humanity as prescribed in the scriptures and so on are preserved. Furthermore, all scientific history and advancement in our civilization(s) will have to be meticulously preserved, and all of it in a non-destructive electronic format. *Remember there are two things distinct about us humans, walking upright and the gift of the language.*

I would like to indulge in a little "poetry" to elaborate the point more. When "Adam" was told to "Get Down" on the planet Earth, he was given three succinct "breaking news" as they say on the Cable Networks. First the news to "drop down" using words in which an order is implied in *plural* as though Adam is not alone (In the Quran in Arabic "Eh-Bitou"). Now is it possible that there was a giant space ship larger than our Earth and may be even bigger than Jupiter, called "The Paradise", and Adam is being told the bad news to get lost? The second bad news being delivered (in the same verse) is that "some of you will be the enemy of the other". So here is poor Adam and the associates being dropped on an unknown planet and being told that not only they have to defend against other life forms but their own will kill each other as well. Finally, a small sliver of hope is left dangling. They are told that on this planet is their abode and all the provisions for a time (meaning it is temporary). The possibility of returning to Utopia is left alive. Now, tens of thousands of years later, we are ready to explore, conquer other planets and become the Bedouins of the space in ships as large as our own planet. We are the "Vicegerent" not only for the "Earth" but any where there is land, the deserts, the valleys, the mountains, anywhere it can be divided and ownership can be established.

We are ready to "Drop Down" on the Martian Land
We are ready to set up camp on the desert sand
Once more we are being appointed the Vicegerent
Mars is no different, it too is inherent

So now, our team lands on the planet Mars and finds vast stretch of the damaged desert (Called Arabian Terra?). There is no flowing water and no agriculture, it is totally barren. The team is looking for two mounds a quarter mile apart. "Abraham's" wife is running between these two mounds; a suckling infant is sitting in the middle on the desert sand and begins to cry wildly rubbing his heels in despair. The rubbing action results in gushing of the water from under the feet of the infant.

The habitation of the Planet Mars is our duty
The Earth is our guide and filled with beauty
The two mounds are similar, the infant is ours
And the desert is bound - to bloom with flowers

Perhaps one more stanza will do it. Mysticism in me requires that I reveal the hidden. So at this stage the surface area of Mars is somewhat close to that of the Pacific Ocean. The word Pacific means something calm and peaceful by nature. In my language it is also "lethargic". ___So I am calling the entire planet Mars to be the "Pacific Desert". This is new vocabulary.___

The Pacific Desert has welcomed us, the wise
Tomorrow, we will fly to Adam's Paradise
The revelation has come down and it is crystal clear
The water is abundant and the thoughts are clearer

Finally, our language, our culture, arts and craft will evolve on the new planet. A thousand years from now, the mission will continue and the new conquerors will lay their lives just as enthusiastically in pursuit of the greatness of their time. They will, however, acknowledge that our efforts today in the 21st century to populate Mars *reduced (forever) the probability of self-destruction of our species in half.* Remember that it is not about the technology in the end. Some random stanzas are penned below.

I am wandering around in a white suit
Pressurized and with very heavy boots
Designer jackets uncomfortable pants
Oxygen flowing in bulky ugly hats

I am dreaming of the happiness in a football field
Dreaming, playing and watching indeed
Screaming of the fans and a hot dog stand
Showing of the ads and playing of the bands

Olympus Mons is so damn high
Even Mount Everest is a bit shy
Everest is a baby, just two feet tall
And the OL-Mons height is six feet tall (Please twist/read the word Mons as Man's)

Get inside, you're running out of air
I know Mom! It's just not fair
Everyone's got this light large backpack
Why can't I get a supersized airsack?

The words "airsack" and "OL-Mons" could be other examples of changing vocabulary in which the word like the "airsack" is an invention, regardless of the constituent words, used as a noun. Here airsack is just a bag that holds larger quantity of air mixture for children playing outside the pressurized living quarters on Mars. Clearly similar words will pile up when going to visit the giant volcano/mountain called "Olympus Mons". This is just child stuff – wait till the grown-ups get involved.

Clearly, the most powerful characteristic of humans is their ability to learn and coin words. "In the beginning was the Word, and the Word was with God........" Does the "word" without the gimmicks of big G and small g (the concept did not exist in old Hebrew, Arabic etc.) could imply that it is our ability to adopt and to coin the words; this is what makes us gods relative to other animals. The fact remains that humans cannot even agree on a proper noun for the "creator" of the universe(s). From the aborigines of Australia calling their god as ATNATU (meaning a being without flaws or more literally a being without anus) to the Jewish word ELOHIM to anything anyone wants, there are gods by the thousands. This shows a level of diversity and a never ending appetite for learning and creation where the language has no boundaries. The mathematical concept of infinity shows that our knowledge will continue to increase, forever, until we self-destruct. And as pointed out earlier, it is this reduction of probability of our self-annihilation that we have embarked on; a mission to colonize the Planets and Mars is only the first step.

We end this chapter by proposing a giant electronic two-way back-up system. Knowledge from the population on Mars records and the systems on Earth are updated and vice-versa. It is clear that without this knowledge, the Martians will have to invent much on their own. This includes the basis of law and so on. The Earthlings will lose as well. We also have to face the fact that someday the Martians will not accept edicts from Earth, unless we can show by example that we have always stood by them and it is the same species working together for self-preservation.

CHAPTER VIII

Birth & Burial on Mars

8.0   Giving Birth on Mars

Setting up any "colony" on Mars will require planning of the ultimate. As described earlier the distances from the home planet are vast, communications very difficult and no local food sources on Mars; everything has to be imported, at least in the beginning. Unless only those folks are allowed to go and settle who cannot give birth (perhaps an option in the beginning); human nature to make love and reproduce is and will remain uncontrollable. In fact, it will be a tragedy and purposeless conquest to not have live birth(s) on Mars and children growing up on the planet, born citizens of the new principality and guardians against the self-destruction of the human race.

We know that the birth of a human child is still a "sacred" duty all over the world, and has been so, for thousands of years. Even in the US, if we care to look at a hospital's departments focusing on the birth of a child, we find, able doctors busy (using in-vitro fertilization and other techniques) helping the women, who are desperate to have a child. We also know that it is not just the "biological urge", laws of many countries have been refined over thousands of years to allow for some legal mechanism (other than say a Christian marriage of one man-woman pair, ignoring the birth of Jesus himself) to have children, as historically vast numbers of humans have perished for any reason; i.e. plagues, wars, feticide, infertility, and so on.

Thus the children born on Mars will be very precious, at least that is how it should be, and every effort will have to be made to make sure that they do not succumb to an untimely death for whatever reasons. The "residential quarters" established on Mars, in the beginning, if they have fertile women, must allow for the possibility of live birth both in the space stations and in these quarters. This means all the necessary medical and other tools to allow for not only the "miracle" to happen, but to make sure that no harm comes to the child even if an evacuation is required either to the space station orbiting the planet, or ultimately to the Earth. The preference, of course, will be for the child to grow with his "peers" on the new planet. Wouldn't it be nice to have twins, pre-planned (the story of Adam's children comes to mind), if at all possible?

a) Of the many experiments to be performed, one should be, to plan for a human birth, first on the space station and then on the surface of the planet Mars. Since the time to reach Mars will likely be around nine months or less, synchronization with the gestation period is possible. Clearly all pre-natal care is required in addition to certain other measures. We will recommend a daily routine in the space station, including an enhanced measure of all the vital signs, affects caused by weightlessness, and suggested remedies for all these adverse effects and possibilities. The development of the baby has to be meticulously monitored and properly measured, ensuring that the health of the mother is not compromised. We know that the buoyancy in the water allows for a feel of weightlessness, thus to some degree the "morsel of flesh" is used to this degree of freedom. Perhaps nature has already prepared women for space ship births, but must be verified. The "baby" should therefore feel less of an affect(s) of weightlessness until the water breaks. Clearly the additional weight of the baby will be easier to support for the mother, but the loss of any mass can also be dangerous. This experiment should ideally be conducted on two or three women (of different races?) to make sure the experiments are not fraught with one type of genetic errors.

b) More important than the experiments suggested in a) above are those concerning the development outside the womb of the mother. Our biggest concern is that the normal "earthlike" growth shall be stinted due mainly to either excessive weight gain to compensate for lack of gravity or the inverse of it; i.e. loss of mass due to lack of hunger for unexplained reasons. The challenge will be to diagnose and take measure to not reverse back in time from evolutionary perspective. Any abnormality must be carefully measured, and remedied, otherwise the entire experiment of colonization process will suffer one impediment after another.

c) The human brain is a labyrinth (or consisting of very intricate passageways) of millions of specialized nerve cells. These cells are interconnected by billions of electrical (and chemical) paths called synapses (or the points at which a nerve impulse passes from one neuron to another). The post-synaptic density or PSD within these synapses are proteins that combine together, forming a molecular machine, which is believed to disrupt synaptic functioning, causing behavioral change and disease. The human evolution has numerous problems to be faced at Mars. This includes lesser gravity, muscle contraction, possible effects on the eyes, ears, and many others phenomena requiring systematic studies on the surface (or well underground) of Mars. Even more crucial will be the mental development and effects of the confined environment, lack of certain types of lighting and electromagnetic waves. The most difficult part of the study has to do with the mental development of a new born child both in the space station(s) as well as on Mars. The tasks include a formal set of physical and mental developmental measurements, including the ability to walk upright without any psychological issues. A number of tests must be carefully designed to measure the development of various physical parts, such as Cerebral Cortex, limbic system, thalamus and the list goes on. We should remember that although the brain is "protected" by the fluid, cavity developments have not been measured in outer space or on Mars with different gravitational structures. There are other aspects of child development that are a great challenge as well, in this context. The plan must call for immediate evacuation back to the Earth, under certain circumstances; certainly if it appears that "speech" related issues have started to develop despite meticulously designed tasks. We are reluctant to suggest any psychological tests other than recommended by child development experts. This will require significant research by the doctors themselves as almost none have any experience of outer space child development issues in the first ten or so years. The most crucial of course are the first two or three years. The child must walk, talk, and behave "normally". It may be that during the "walking training" a new apparatus or equipment will be required. ***The child must not lose the upright walking capability. It is the one thing that makes us who we are; and if it turns out to be a real problem then a new criterion has to establish until all problems are resolved and perfection achieved.*** In the meantime the children are transported and develop on the Earth, while allowing for the actual birth and possibly the first few months in space beyond birth.

8.1 Tradition of Burial

It was told by my grandmother that I was barely two years old when my father passed away. As if this was not enough, my mother also left this mortal world when I was just turning seven. The neighborhood almost had no children of my age. I had no real friends. My grandmother and I, and the little house, sometimes felt haunted. There was a very nice old couple that had no child of their own, couple of blocks away, and my grandmother used to take me to their house at around 7:00

pm almost every day for formal lessons, not taught in school; things about the past, and how it affects the future, things about the stars, learning about another language, and so on. That was the way of my childhood. This man was a homeopathy doctor and had so much knowledge, and I, a little child, used to get lost thinking about the things he would tell me. In hot summer, in my little home's open court yard, under the stars, I used to think all night dreaming that someday the humanity will have to leave this planet and face the difficulties where parents would die, and children will be left alone to fend for themselves, and then I would dream of cemeteries of the astronauts.

One day as we were in our beds, my grandmother told me a story. A long time ago there was a Holy Man full of wisdom and knowledge. A poor child who was an orphan wanted to learn a lot, be a scholar; back then there were no formal schools and every child was working in the fields and the farms helping their parents. One day the child just walked over to the house of this Holy Man, and asked, if he could learn about the Almighty. The learned man said "he was very busy" and that he had no time for him. The child became very sad. The man looking at the sad face of the child said "well I have a little time I could give you while I am commuting to the King's court. You can ride with me but you will have to come back on your own". The child happily agreed. So every morning he will go with the teacher and come back this very long way exhausted and hungry. The holy Man observed the behavior of the child and knew the student was not only serious but brilliant as well; he had the qualities reminding him of his own childhood. Seeing his enthusiasm and dedication, the Holy Man hugged the child and promised he would teach him all that he knew. The child grew up to be a great teacher in his own right; his tomb is built beside his teacher's; and even today people come here from faraway places to visit and to learn. My grandmother finished the story and afterwards fell asleep leaving me to my thoughts. Will I be buried in a tomb when I die? Will people come to visit my grave? And then I started thinking about the Taj-Mahal and the pyramids, tombs of the yesteryears. Many years later, I would learn that the pyramids are pointing in the direction of the stars of the Orion Belt, and to my surprise, the great Taj-Mahal is built, far more, to glorify the

Almighty than mortal love; the inscriptions of the Quranic chapters and verses speak for themselves.

So from the dawn of civilization, man has paid attention to the burial duties, and in my humble opinion, the origins of the thoughts of an organized religion, came from pondering on the rites and rituals following the death of someone they thought could never leave them (or could never be replaced). This is especially true when a great man dies, and people refuse to acknowledge the facts, succession disputes breakout and some even become immortal, never to "die" (i.e. Hindu gods or as atheists have asked, is Jesus such a person?). There is a famous incident in the Muslim history; when Mohammad died, the would-be second caliph, Omar, became so emotional he took his sword out and threatened to smite anyone who said "Mohammad is dead", as it is a "sin" to say so. However, the man who succeeded Mohammad named Abu-Bakr, the first Caliph in Islamic *history* knew better. He stood on the pulpit and made a powerful speech: Lo! As for him who worshipped Mohammad, Muhammad is dead, but as far him who worshipped God, God is alive and dies not. He then recited a verse of the Quran. (III-144) "Mohammad is but a messenger, and messengers (the like of whom) have passed away before him. Will it be that when he dieth or is slain, ye will turn back on your heels? He who turneth back doeth no hurt to God, and God will reward the thankful". This speech and the quote from the Quran by the first Caliph took the wind out of any other ideas on the subject, and an orderly transition of succession proceeded.

Therefore, it is prudent for us to ask what will be our stance when someone will pass away from the scene on Mars and there are no relatives or laws on the planet. What if the bodies are pulverized, as in the Challenger or Columbia accidents or not found? Will some of the rituals be performed and how? What if the body is found but lack of water and other agents become an issue for burial. There are no funeral homes and staff to "prepare" the dead bodies. While we may think of these as not so important issues, they suddenly become very important, emotional and even rancorous on Mars; and if suspicions of foul play exist, it might lead to bloodshed. We will be remiss, if we did not admit a tendency among the humans, that of Cannibalism.

I don't necessarily mean the actual act of eating of their dead by the same species; but the act itself is alive and well, be it for "good" purposes namely organ transplant or car parts. In the beginning there may exist occasions when our animalism shall be put to the ultimate strain. Some rules of triage are therefore essential and must be part of the "training manual" for the astronauts and all potential immigrants to the new colony.

8.2 Formation of Interplanetary Burial Commission

There are many things and organizations to be formed that are the first and unique in their mission. Death is part of the life we have known, and will continue to be so for a long time, at least until paradise (life forever) has been discovered? We therefore need a commission to sort out the details of the disposal of human remains on an interplanetary scale. It is clear that there will be some very "rich" people on different planets who would want their loved ones returned back to the home planet. There will be others who will prefer to be "buried" in their new surroundings. Primarily the job of the commission will be to arrive at "rules" to avoid contamination throughout the "system" whatever the definitions may be. *One of the main issues for Mars is to make it habitable; the commission has to keep that in mind while arriving at these rules of burial. Those who die on Mars perhaps will donate their bodies for that purpose. This will be for a higher purpose to increase chances of human survival as a species.* We know that on the Earth, we have two primary methods of disposal of dead bodies. First, there is the burial, and the second being the incineration. The Christian "burial" has one drawback in the sense of the grave site taken forever and thus not usable by anyone else. The Islamic system, only marginally better, allows for the administration to use the same site after a few dozen or so years, and thus the cemetery remains in use without increasing its size significantly "forever". Clearly, now that a very good mechanism is available where little or no residue is left, incineration is becoming the preferred method, except among the Muslims and a majority of Christians. The commission will obviously require all, to precisely detail their wishes. In any case detailed rules are the function of such a commission. My personal suggestions are detailed below, although religious experts must also be consulted and so on.

a) Those who would like to be buried in the "traditional ways", their remains shall be left in the cemeteries created behind the places of worship (Chapter VI). Since the surface of Mars is extremely cold and any bacteria will not survive, initial burial shall be very simple and it will be to leave the body in a surface chamber properly disposed in a dignified manner, provided there is sufficient time. Saving the alive will be, always, the highest priority as retrieval of a naturally preserved body is always possible later (much like it is on earth; Mars is a freezer anyways). As there may be no time to go through the rituals, those can be performed on Earth probably as a group, if more than one or as individuals. Various religious experts can give their views; in my mind this is allowed even now.

b) Those who would like their bodies incinerated will have to wait until such mechanisms can be set up on Mars. Those planning to set up colonies obviously will set up the necessary equipment at the appropriate time.

c) It is clear that as time passes, all of the facilities will have to be set up. This includes a highly organized and efficient system of recovering a "person" both in the physical but also in the mental sense. The vastness of the wilderness will have to be respected. It is quite possible to lose one's senses in the middle of the game and a behavior observed that had never been seen before. The physical recovery is perhaps easier, from the point of an accident to the nearest facility (i.e. a hospital?). This facility may be underground, above ground in the space station or somewhere in between. The ultimate destination might actually be on Earth, at least in the earlier days of colonization.

d) Finally, I am unable to predict the future, but a sense of privacy must be preserved for those who do not wish to be exposed for any reason whatsoever. All rituals must be performed with utmost respect for the privacy expected.

8.2 Rules of Succession

The structure of the Local Government shall be discussed in Chapter IX. Here we end this chapter by simply asserting that strict rules of job successions must be followed. In other words no one on Mars shall be allowed to "settle" until they have completed a formal survival course, a code of ethics, and signed a covenant to help other humans until disabled. Once a person dies, his assigned tasks must be transferred to the successor. Simultaneously, no one will be asked to do something that they don't have an aptitude to complete.

We are hopeful that humanity will not make the same mistakes we have been making on the planet Earth. To help in the birth of a child on Mars shall be a sacred duty. The burial and succession will be simple, dignified, and without fanfare, including for those at the highest "official ranks".

CHAPTER IX

Initial Structure of the Government on Mars

9.1   The Early Exploration Period

I was ambivalent about writing this chapter simply because I don't know much about the structure of any government. However, that may be an advantage, as perhaps a new formalism must be at least looked at. In the beginning when the exploration phase is underway, thinking about a structure of the government on Mars is silly. The authorities on Earth will obviously control the entire phase. That includes initial travel plans, spaceship trajectories, and ultimately landing on the planet Mars. The commander of the ship will have all the authority and in case of an emergency, a mishap, a complete plan of survival, if survival is at all possible will be the responsibility of the commander. If the commander is killed, a succession order will be instantaneously implemented. The safety, security, and the well-being of all will then be the responsibility of the new commander. The destruction of two space shuttles, Challenger headed for the International Space Station (ISS), and Columbia coming back from it, both very near the Earth have taught us a lot. Many have given their lives to push the space program forward.

An area of safety that bothers me as inadequately planned is that of mental competence. Since all the space travel has been relatively short, it is unclear on how the commander is supposed to behave in conditions that none of us have observed or planned for, and vastness of the space when nothing is near us and when the *dormant whispers* shall come alive, their diagnosis, and cure are real concerns. We really don't have any data on how the loneliness of space, lack of certain feelings, (not just lack of gravity) will affect us. Even at ISS we are too close to the home planet. It is even unclear to me if Mr. Neil Armstrong is really a "normal" person after a successful mission to Moon. The folks who have landed on the Moon are still a mystery to most of us, and it is only 250,000 miles or so away. If we take a journey, that is 40-50 million miles away, and plan to come back we have to actually plan for the commander becoming sick and unable to discharge his/her duties, for both physical and mental illnesses. For example, if the vision of the commander begins to deteriorate due to prolonged lack of gravity, space motion sickness, and the behavior is suspect due to excessive flow of blood in the upper part of the body, massive loss of muscle and tissue, bird leg syndrome, and so on. But most importantly, *loss of confidence due to prolonged sickness must be detectable and curable during flight; that is crucial.* A succession regime must include the solution of these issues before it can be implemented.

One last point in this early exploration period that explicitly needs to be conveyed is the selection of people who are culturally homogeneous. We are not talking about race here. However, language is the most important aspect of this early era. Homogeneity and understanding of the idiosyncrasies of the language(s) are essential in early detection of any psychological issues facing the explorers. Thus if there are only three astronauts, they must all speak English or French or Urdu or whatever. It will not be wise that the first one is native English, the second native French and the third native Urdu speaker(s), and only understand and communicate with each other in "scientific" style English. While they may ultimately recognize each other's symptoms, it might be too late. These conditions are, however, *not* mission specific.

9.2 The Medium Exploration Period (200-400) Years after the First Landing

I do not wish to assume that all has gone well in the first two hundred years since the first successful mission to Mars. During this period, the likelihood of man being able to sustain life on Mars, *based upon the resources on Mars*, is slim to nothing. Most likely almost all the sustenance comes from Earth. The number of people actually living on the space stations and on Mars is approximately twenty thousand at this stage of the exploration. However, during this period, people have begun to see a small growth of food and other supplies on Mars, enough to sustain a few thousand people. It is at this stage that an urge to "break relations" with Earth might come to Martian residents for whatever injustices they perceive. The thought of sending spaceships to quell the rebellion on Mars, or to impose economic sanctions on the colony, will be the most stupid actions, the earthlings can take.

Therefore, a system of government must be anticipated for this period that is just, gives the new colony maximum flexibility to govern their lives, and lays the foundation of a peaceful relationship with the mother planet. First, a leadership structure must be envisioned, and feedback required in the first two hundred years, that continues to strengthen the relationship with an institution that represents Earth rather than 200 governments singing their own tunes and confusing the hell out of the residents on the new planet, and even resulting in a war on Earth. They, the Martians, should not be the pawns of our insanities on Earth. At the moment we have no such "institution". Whoever contributes money and the technology calls the shots, and the contribution of the other is always considered less than our own. But we have a lot of time (100 years to sort it all out?). I am somewhat skeptical about the governance structure of the United Nations, but if important changes can be made, for space exploration, they should be acceptable to us all, at least the UN should be looked at as a viable institution. Ideally, a requirements based structure should be established. Remember, that we might design spaceships that are so large that a dwarf planet, a moon, or a small satellite, may have less people on it than a spaceship. So the idea is that, while they are busy in the service of "humanity", they are not running fiefdoms against each other, at this early stage of the exploration.

9.3 The Longer Term View of the Government on Mars

I personally do not view Democracy as an ideal form of government, at least in the context of space exploration, as it has too many flaws to list here. I am also not advocating even a benign dictatorship. Rather a new form of government that is based upon the "Principle of Continuous Improvement" (PCI) discussed elsewhere. The form of this government is critical, as a self-destruction of humanity is at stake. We cannot propose a "federation" type of government or any other. Here on the planet Earth, we have been unable to control wars, the biggest source of killings. Now, that Antarctica might have resources, we will see how, it can be used as a model for Mars? I am somewhat skeptical. In general, the "Constitution" must allow, for a quick change in the implementation of any "statute" (as the devil is in detail) without the usual litigations, in a preventative sense. If there is scientific evidence that the statute must either be improved or made part of a larger statute, it will be implemented just as quickly.

For example, the fundamental premise and the reason for the existence of "murder" statutes will be to make sure that there is never any murder or blood shed on the new planet. The only exceptions and reasons for death will be natural or a genuine or unplanned accident. If *murder* happens, then heads must roll in the government ranks as well.. In other words, the fact that murder has occurred, is just as much the fault of the authorities, and sufficient preventative measures had not been taken. The preventative measures can be mental or physical, lack of training, lack of whatever. Put, in another way, the "blame" for murder must be shared by the authorities, and not just that of the "criminal". Note that we will not allow the creation of a police state. We are not interested in arguments that have to do with 'passions" based murders and so on. The fact is that self-control must be inculcated from childhood and proper medical inventions and help demanded (remember we are talking about a power structure hundreds of years from now). Even in the United States where murder rates are high relative to many civilized countries and capital punishment advocates abound, the author lived in a town of more than two thousand people where no felony (let alone murder) had been committed in the last one hundred years. The author wishes the town to keep that record and so will not mention the name. In future, we will look to PCI and use it to modify the "statutes" so that murder will never happen again. There will be no "punishment" for the "criminal" only "genuine medical reformation". All psychological disorders must be prevented and treated humanely and with due respect.

While we are not all knowing, PCI should help us in continually modifying the structure of the government on Mars and other colonies. The objective shall be to serve the people on Mars, and advance the cause of exploration with the goal of reducing the probability of self-destruction of humanity on a planetary level. This means that humanity will survive if the Earth is swallowed by the Sun. Ultimately this means that humanity will survive if the entire solar system collapses. This means that humanity has achieved its goal of reaching the "paradise" never to face complete extinction for any cosmos related accidents or self-destruction.

APPENDIX I

Scripture references from the Old Testament, the New Testament, and the Quran have been mentioned throughout this book. We are not going to repeat those quotations here. In this appendix we are providing specific references for those interested in the data contained in the original Arabic text of the Quran. The data consists of all those locations (chapter and verse) where the words "Earth", "Heavens" or both have been used. This is to the best of our knowledge. If someone can find additional locations, we will be glad to acknowledge them and add to the spreadsheet. For me, personally, it was fascinating that the Quran has two-hundred and twenty locations where _both_ of the words "Earth, and Heavens" have been used in the same verse. Many examples have been cited in the text of this book already. This appendix should allow further research at the appropriate levels and theological directions of continuing scholarship (please also refer to footnote 3 in Chapter II). The trans-literation here is generally accepted as standard.

Sipara	Surah	Ayat #	Ard	Samaa	Both	Comment(s)
Juz 1	Al-Fatiha					
	Baqarah	11	*			
		19		*		
		22			*	

		27	*			
		29			*	
		30	*			
		33			*	
		36	*			**Planetary Stay is temporary**
		59		*		**Habitation?**
		60	*			
		61	*			
		107			*	
		116			*	
		117			*	
Juz 2		144		*		
		164			*	
		168	*			
		205	*			
		251	*			
Juz 3		255			*	
		267	*			
		273	*			
		284			*	
	Aal-e-Imran	5			*	
		29			*	
		83			*	
		91	*			
Juz 4		109			*	
		129			*	
		133			*	**Paradise very large**
		137	*			
		156	*			
		180			*	
		189			*	
		190			*	
		191			*	

	Nisaa					
Juz 5		42	*			
		97	*			
		100	*			
		101	*			
		126			*	
		131			*	
		132			*	
Juz 6		153		*		
		170			*	
		171			*	
	Ma'idah	17			*	
		18			*	
		21	*			
		26	*			
		31	*			
		32	*			
		33	*			
		36	*			
		40			*	
		64	*			
Juz 7		97			*	
		106	*			
		112		*		
		114		*		
		120			*	
	Anaam	1			*	
		3			*	
		6			*	
		11	*			
		12			*	
		14			*	
		35			*	
		38	*			
		59	*			

		71	*			
		73		*		
		75			*	
		79			*	
		99		*		
		101			*	
Juz 8		116	*			
		125		*		
		165	*			
	Aa'raaf	10	*			
		24	*			
		54			*	
		56	*			
		73	*			
		74	*			
		85	*			
Juz 9		96			*	
		100	*			
		110	*			
		127	*			
		128	*			
		129	*			
		137	*			
		146	*			
		158			*	
		162		*		
		168	*			
		176	*			
		185			*	
		187			*	
	Anfaal	11		*		
		26	*			
Juz 10		63	*			
		67	*			
		73	*			

Juz	Surah	Verse				
	Taubah	2	*			
		25	*			
		36			*	
		38	*			
		74	*			
Juz 11		116			*	
		118	*			
	Yunus	3			*	
		6			*	
		14	*			
		18			*	
		23	*			
		24	*			
		31			*	
		54	*			
		55			*	
		61			*	
		66			*	
		68			*	
		78	*			
		83	*			
		99	*			
		101			*	
Juz 12	Hud	6	*			
		7			*	**7 day Creation and throne over water**
		20	*			
		44			*	
		52		*		
		61	*			
		64	*			
		85	*			
		107			*	
		108			*	

		116	*			
		123			*	
	Yusuf	21	*			
Juz 13		56	*			
		73	*			
		80	*			
		101			*	
		105			*	
		109	*			
	Raad	2		*		
		3	*			
		4	*			
		15			*	
		16			*	
		17			*	
		18	*			
		25	*			
		31	*			
		33	*			
		41	*			
	Ibraheem	2			*	
		8	*			
		10			*	
		14	*			
		19			*	
		24		*		
		26	*			
		32			*	
		38			*	
		48			*	
Juz 14	Hijr	16		*		
		19	*			
		22		*		Seed fertilization thru winds
		39	*			

Juz	Surah	Number				
		85			*	
	Nahl	3			*	
		10		*		
		13	*			
		15	*			
		36	*			
		45	*			
		49			*	
		52			*	
		65			*	
		73			*	
		77			*	
		79		*		
Juz 15	Bani Israeel	4	*			
		37	*			
		44			*	
		55			*	
		76	*			
		90	*			
		92		*		
		93		*		
		95			*	
		99			*	
		102			*	
		103	*			
		104	*			
		110		*		
	Kahf	7	*			
		14			*	
		26			*	
		40		*		
		45			*	
		47	*			
		51			*	
Juz 16		84	*			

Juz	Surah	Verse				
		94	*			
	Maryam	40	*			
		65			*	
		90			*	
		93			*	
	Taha	4			*	
		6			*	
		53			*	
Juz 17	Ambia	4			*	
		16			*	
		19			*	
		21	*			
		30			*	Heaven/ Earth as one piece & evolution recorded
		31	*			
		32		*		
		56			*	
		81	*			
		104		*		
		105	*			
	Hajj	5	*			
		15		*		
		18			*	
		31		*		
		41	*			
		46	*			
		63			*	
		64			*	
		65			*	
		70			*	
Juz 18	Mu'minoon	18			*	
		71			*	
		79	*			

		84	*		
		86		*	
		112	*		
	Noor	35			*
		41			*
		42			*
		43		*	
		55	*		
		57	*		
		64			*
	Furqaan	2			*
		6			*
Juz 19		25		*	
		48		*	
		59			*
		61		*	
		63	*		
	Shu'araa	4		*	
		7	*		
		24			*
		35	*		
		152	*		
		183	*		
		187		*	
	Namal	25			*
		48	*		
Juz 20		60			*
		61	*		
		62	*		
		64			*
		65			*
		69	*		
		75			*	Nothing hidden in the heavens/ earth

		82	*			
		87			*	
	Qasas	4	*			
		5	*			
		6	*			
		19	*			
		39	*			
		57	*			
		77	*			
		81	*			
		83	*			
	Ankaboot	20	*			
		22			*	
		34		*		
		36	*			
		39	*			
		40	*			
		44			*	
Juz 21		52			*	
		61			*	
		63			*	
	Room	3	*			
		8			*	
		9	*			
		18			*	
		19	*			
		22			*	
		24			*	
		25			*	
		26			*	
		27			*	
		42	*			
		48		*		
		50	*			
	Luqmaan	10			*	

		16			*	
		18	*			
		20			*	Heavens/ Earth at our service
		25			*	
		26			*	
		27	*			
	Sajdah	4			*	
		5			*	1 day = 1000 years
		10	*			
		27	*			
	Ahzaab	27	*			
Juz 22		72			*	
	Saba	1			*	
		2			*	
		3			*	
		9			*	
		14	*			
		22			*	
		24			*	
	Faatir	1			*	
		3			*	
		9	*			
		27		*		
		38			*	
		39	*			
		40			*	
		41			*	
		43	*			
		44			*	
	Yaseen					
Juz 23		28		*		
		33	*			
		36	*			

		81			*	
	Saafaat	5			*	
		6		*		Lower planets
	Saad	10			*	
		26	*			
		27			*	
		28	*			
		66			*	
	Zumar	5			*	
		10	*			
		21			*	
Juz 24		38			*	
		44			*	
		46			*	
		47	*			
		63			*	
		67			*	
		68			*	
		69	*			
	Mu'min	13		*		
		21	*			
		26	*			
		29	*			
		37		*		
		57			*	
		64			*	
		75	*			
		82	*			
	Hameem Sajdah	9	*			
		11			*	
		12		*		
		15	*			
		39	*			
Juz 25	Shuraa	4			*	
		5			*	

Juz	Surah				
		11			*	
		12			*	
		27	*			
		29			*	
		31	*			
		42	*			
		49			*	
		53			*	
	Zukhruf	9			*	
		10	*			
		11		*		
		82			*	
		84			*	
		85			*	
	Dukhaan	7			*	
		10		*		
		29			*	
		38			*	
	Jaathiya	3			*	
		5			*	
		13			*	
		22			*	
		27			*	
		36			*	
		37			*	
Juz 26	Ahqaaf	3			*	
		4			*	
		20	*			
		32	*			
		33			*	
	Mohammad	10	*			
		22	*			
	Fatah	4			*	
		7			*	
		14			*	

Juz	Surah	Verse				
	Hujraat	16			*	
		18			*	
	Qaaf	4	*			
		6		*		
		7	*			
		9		*		
		38			*	
		44	*			
	Azzaariyaat	7		*		
		20	*			
		22		*		
		23			*	
Juz 27		47		*		**Universe is Expanding?**
		48	*			
	Toor	9		*		
		36			*	
		44		*		
	Najam	26		*		
		31			*	
		32	*			
	Qamar	11		*		
		12	*			
	Rahmaan	7		*		
		10	*			
		29			*	
		33			*	
		37		*		
	Waqiah	4	*			
	Hadeed	1			*	
		2			*	
		4			*	
		5			*	
		10			*	
		17	*			
		21			*	

		22	*			
Juz 28	Mujadilah	7			*	
	Hashar	1			*	
		24			*	
	Mumtahina					
	Saff	1			*	
	Jum'ah	1			*	
		10	*			
	Munafiqoon	7			*	
	Taqhabun	1			*	
		3			*	
		4			*	
	Talaaq	12			*	**Does not mean 7 Earths**
	Tahreem					
Juz 29	Mulk	5		*		
		15	*			
		16			*	
		17		*		
		24	*			
	Qalam					
	Haaqqah	14	*			
		16		*		
	Ma'arij	14	*			
	Nooh	11		*		
		15		*		
		17	*			
		19	*			
		26	*			
	Jinn	8		*		
		10	*			
		12	*			
	Muzzamil	14	*			
		18		*		
		20	*			

Juz	Surah	Verse				
	Muddathir					
	Qiyaamah					
	Dahr					
	Mursalaat	9		*		
		25	*			
Juz 30	Naba	6	*			
		19		*		
		37			*	
	Naziaat	27		*		
		30	*			
	Abasa	26	*			
	Takweer	11		*		
	Infitaar	1		*		
	Mutaffifeen					
	Inshiqaaq	1		*		**Planets mentioned in verse II**
		3	*			
	Burooj	1		*		
		9			*	
	Taariq	1		*		
		11		*		
		12	*			
	A'ala					
	Ghashiya	18		*		
		20	*			
	Fajar	21	*			
	Balad					
	Shams	5		*		
		6	*			
	Layl					
	Dhuha					
	Inshirah					
	Teen					
	Alaq					

	Qadr					
	Bayyinah					
	Zilzaal	1	*			
		2	*			
	Aadiyaat				NR
	Qariat				NR
	Takathur				NR
	Asar				NR
	Humaza				NR
	Feel				NR
	Quraish				NR
	Ma'oon				NR
	Kauther				NR
	Kaafiroon				NR
	Nasr				NR
	Lahb				NR
	Ikhlaas				NR
	Al-falaq				NR
	Naas				NR

APPENDIX II

In this appendix, details of the proposal (new Day) the "ND" and Evolutionary Level "EL" are examined from a religious, perspective as religions have aided in the measurement of time. An attempt shall be made to prove that the changes being proposed are consistent with the scriptures.

II.1   Measuring Time and the Religious Thought

Jewish, Christian, and Islamic scholars, based upon the acceptance of the concept of God and the scriptures, have explained the fundamentals already and I am not the ultimate authority. Here *we shall accept the hypothesis of the existence of God for the purposes of its relevance to the measurement of time* – as time has been an integral part of the religious life – i.e. birthdays, death anniversaries, and uncountable events in between. According to these beliefs, both the physical and spiritual Universe(s) were created by God Almighty. He is the one entity, which spans all universes, physical or otherwise. He is independent of relative time and "Absolute" in nature. This simply means that the belief dictates that He is the only real constant in all of the Universes and unlike our Time neither relative nor measurable. The precise mechanism utilized by Him to create these universes remains unexplained, or what has been religiously (or scientifically) theorized to date. The only religious explanation of time is the number of days as six in the Bible or six "periods" as in the Quran and the process of creation is simply the cause and effect. The Almighty said, "be" and "it was".

My own religious thoughts are that He first created the *principles*, which then became the *basis* of all creation. *This is important as it implies an abstract formulation first.* If we accept this postulate, then Science and Religion are one and the same, only different in emphasis. Science emphasizes more on the origins of these fundamental *principles* and the precise physical mechanism. This leads to physical experimentation and the probe into many details that are still unclear. The religious emphasis on the other hand is primarily concerned with the creation(s) themselves and their behaviors. The self-aware among the creation have thus looked for a *"Guiding Light"*. The use of the words *Guiding Light* is simply my way of saying that all fundamental religious *principles* can be *approximated* for the *self-aware* into a *relative legal framework*. This legal framework has to guide us even as we invent those scientific laws, and the hope is, without self-destructing ourselves. The inertia created by the religious institutions, even at the expense of human suffering, comes from that guiding light (or the perceived *legal framework*) and has a calming effect on our desire to move unchecked. When all aspects of a problem have been understood, the guiding light moves us forward. Thus if we are slow to begin biological cloning for religious and ethical reasons, it comes from that inertia embodied in self-preservation. It comes from that *guiding light* (or the perceived *legal framework*). Similarly, the physical interpretations of the *Causality* and the *Uncertainty* principles are valid both in science and religion and the only real constant by definition is the Almighty. He may have embedded a few *universal constants (i.e. the Plank's constant)* in nature to guide us into accepting His being the Absolute. After all we have invented the mathematical concepts, and mathematics is abstract. Let's move the argument a bit further, as anything abstract is not necessarily going to be or part of the *guiding light*.

II.2   Abstract Notion of His Image

A thought that has some validity in all higher religions is that the Almighty has created human beings in His image and given us some of His qualities. What does this mean if He is never changing and absolute? Again, my thoughts are that the result of cumulative *intellect* of all humans and derivative knowledge, integrated over infinite period of time, is the only plausible object that can even come close to being approximated with His absolute knowledge. Please note that the individual does not survive in the physical body forever - however the body of knowledge created by the individual has the possibility to survive long after one has physically perished "forever". We being the self-aware creation recognize ourselves. We invent, remember, and classify the known for posterity. The next generation builds upon that intellect and knowledge and adds to its base. This goes on forever and ever. In thousands or perhaps trillions of years, we may fully comprehend Him, as we begin to approximate this knowledge base spanning the universes. Whether that will happen remains a big question, as we *appear* to be self-destructive as well. In mathematics, we could therefore say:

$$\text{Absolute Knowledge} \geq \int_{-\infty}^{+\infty} \text{Relative Knowledge} \qquad \text{(II. 1)}$$

Here Absolute Knowledge means the "Knowledge of the Almighty" while Relative Knowledge means human (or non-human) or measurable knowledge in its totality. Thus

Relative Knowledge $\leq \Sigma$ Human Knowledge + Other Knowledge

Human Knowledge = CI

Where C = Coefficient of Absorption (averaged over the lifetime of the species – note that a specific generation may lose some knowledge but the species may still have a net gain – what can be absorbed as a whole for the humans as species). And where I= Human Intellect.

I do not wish to go into further details about "Other Knowledge" in any future discovery of the inter-intra-terrestrial intelligence, yet which by definition is not the Absolute. However, these formulae have to be kept in mind and lead to what we have referred to as the Principle of Continuous Improvement.

II.3   The Theology of Higher Religions

We are not interested in the esoteric differences among the three great religions, as the Jews and Muslims clearly believe in the same God - the Christians get there as well with a tangential delay. While the Jewish and Islamic laws are far more specific, the Christian laws are not (some still believe that Jesus came to annul the law – hence eating of the pork, circumcision, etc.). For two reasons, we shall concentrate on references from the Islamic Scriptures, to make the point on the matter at hand, namely measuring Time. First, Islam is the latest and the newest in the evolution of these great religions, and thus (hopefully) more scientifically evolved. Secondly, it is my personal choice as I am less familiar of the others – but to the extent, possible, a similarly strong argument shall be mounted from all three. The Arabic word *Ijtihad (not to be confused with Jihad akin to terror in the West)* in its most widely understood meanings simply refers to the *effort* resulting in the establishment of a process which, by using the *Guiding Light* determines the most judicious path of solving a specific problem, without violating any of the principles stated in the religions. The questions we are raising in this book are serious and those that we believe shall affect the nations for hundreds of years to come, if not eternally. What we are suggesting therefore is doing Ijtihad in the lingo of the Islamic scholars. My Ijtihad is not binding on anyone because part of the Ijtihad process is to arrive at a consensus among the nations. *In my opinion however, we must continuously evaluate and re-evaluate our posture on any and all matters necessary to stay ahead, and lead the world at large.* We are therefore, asking those who think of themselves as the *leaders of all these great religions* to at least give the questions, if not the answers, a few solitary moments of deep thought. If we acknowledge that there might just be a better "religious option" for us, then it will be our religious duty to carry out this process. It may be that after careful considerations, and experimentations, nations still arrive at the same conclusion as today, and keep following the current Calendar(s). Let us proceed with the Islamic thought in researching out this religious *legislation*.

1. The first step in Ijtihad is to present the problem in a concise manner, which we have done (in the body of the book - chapter III).

2. The next step is to propose a solution, which we have done (Chapter III section 3.6 and beyond).

3. The third step is to justify the solution from the "scripture" in following the process. We shall do this now. In a nutshell, the Quran forbids intercalation (IX:37) – *"Verily the transposing is an addition to unbelief; The Unbelievers are led to wrong thereby: for they make it lawful one year, and forbidden another year, in order to adjust the number of months forbidden by God and make such forbidden ones lawful. The evil of their course seems pleasing to them, but God guideth not those who reject faith"*. The pagans used to change the months around (subtract or add days even months in a specific year) especially whenever it suited them to justify wars in the so called "sacred months" - While the believers respected the sacred months and were at a disadvantage in a sudden onslaught of war. The pagans did this even though it was forbidden by their own belief. It should be noted that in those times (long before Pope Gregory) there was no precise way to count and fixate months except using the Lunar Motion. So every so many years, the intercalation by one month brought the year roughly equal to the Solar Year. It should also be noted that in the previous verse, the Quran declares the months to be "ordained" as 12 at the creation of the Earth and the Heavens (including the Sun and the Moon). Pope Gregory, and everyone else, continued the practice of keeping the months to be 12 while trying to adjust the months to contain 28, 29, 30, and 31 days, and introducing the concept of Leap Year – a mild form of fixed intercalation. The Quran clearly wants us to

stop the intercalation – and might I suggest to the degree that observation and Uncertainty Principles would allow us. It should be noted that intercalation has NOT stopped in either Lunar or the Solar Calendars. The Israeli attempt to fix the Lunar Calendar even though remarkable, still has not fully resolved the issue of intercalation. The reason is that until now, it was not possible – so we did the best we could from the perspective of experimentally counting time as humanity marched on. Thus what the religions really want us to do, is to "continually count time" and ignore the celestial motions if we are to completely stop this intercalation which is not only the position that Quran takes but also taken by Pope Gregory and the modern Jews – a position also required by science. We have thus proven from the Quran that our methodology is in its accordance and certainly not against it.

4. The next and final step in the Ijtihad process is to seek guidance from any explanations contained in the teachings called "Hadith" observations and guidance given by Muhammad. *Let us begin very simply by noting that the most important principle in this methodology is to recognize that unless something is explicitly forbidden, it is most likely allowed. It is not so that all things are forbidden and only a few are allowed.* Eating of Pork is explicitly forbidden for Jews and Muslims, and so on. If we accept this principle of "jurisprudence" (or the legal basis), then we can safely say that there is nothing in the teachings of the Prophet(s) that violates the "Principle of Continuous Improvement". There is also nothing that prohibits the measurement of time using a clock, or a clock like device. What is being proposed is a continuous summation of time from its most fundamental unit using the human being's biological clock. Thus any old style measurement is a subset of the new system. The lunar month can still be preserved for posterity, much the same way that Roman Numerals are. If, however, we want to go deeper, the fact that the founder of Islam (and the Jews before him) chose a lunar month is actually in favor of the argument, we are proposing. *The lunar year revolves around*

the solar year. It thus sets the precedence for us to generalize this further and say the "Human Day" also revolves around the solar day. If the time is measured in this way we need not count it artificially using the Moon and the Sun. These heavenly bodies are nothing but signs of the Almighty for us to believe in Him. Humans, being the *most elevated among His signs,* need not measure their time using lower levels of his signs, when the signals are contained within. *It may be a surprise to the non-Muslim, but the founder of Islam was asked about how five times per day prayers will be conducted, if the day were to* **become longer than 24 hours,** *as there was not any clock to rely on during those days. The question was asked in connection with a character to come, in future named Dajjal, when some days will be longer than 2 or 3 solar days. He responded that we will calculate time.*

We have thus met all the requirements imposed by Ijtihad (in providing the solution to the problem and using The Principle of Continuous Improvement).

APPENDIX III

The author has previously suggested a triad system of energy production, without the use of any fossil fuels for Earth. The diagram is being modified for Mars. Clearly, a "wind/storm system pattern(s)" for Mars are required study.

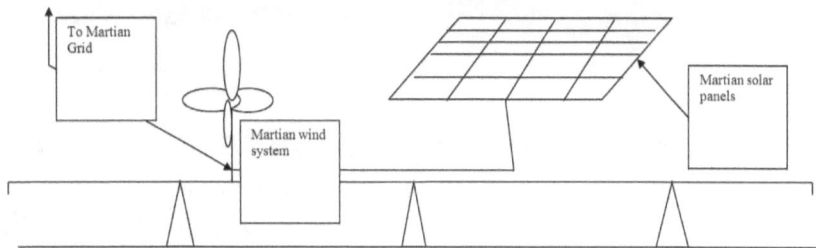

Please note that a ring *on the surface* of Mars of about **5 miles by 13000** miles along the "equator" will provide sufficient electricity for the planet to not only help it warm up but also for every imaginable use. The Sun shines longer on Mars, and the storms may last a long time, although a new wind turbine design will be required.

Three Interesting Pictures

The author has selected three pictures to show that NASA scientists have attempted to keep a historical connection with the Earth in the naming of various Mars "neighborhoods". These are somewhat obscure pictures, and displayed here to generate habitation interest on Mars by ordinary folks the world over.

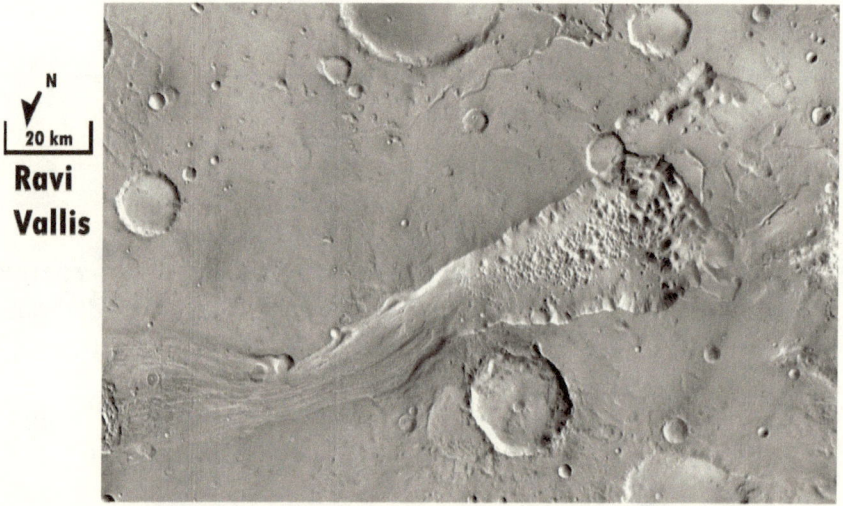

Picture I

Ravi Vallis on Mars: Named after the Pakistani/Indian River "Ravi" (flows through the fertile plains of the greater Punjab province and ultimately joins the historic Indus River, known for Indus Valley civilization. The river Ravi originates in the towering mountains of Kashmir). NASA file in Public domain. The branched channel(s) seen by Viking from orbit suggests that it rained on Mars in the past. Ravi Vallis is located in the **Margaritifer Sinus quadrangle on Mars**.

Picture II

A recent picture – A panoramic view of the Eagle Crater on Mars (shows outcroppings of water origin, again located in **Margaritifer Sinus quadrangle). NASA file photo in Public Domain.**

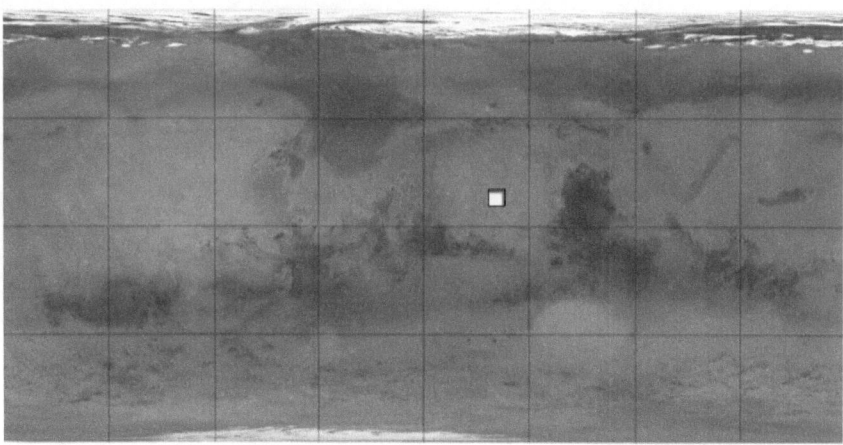

Picture III

Arabian Terra at E19.79°N 30°E - The small white rectangle, inserted by the author, is roughly the proposed location of the "Worship Corridor" on Mars' (please refer to Chapter VI). The Coordinates provided are based upon a two dimensional NASA map.